A Practical Guide to Chemical Spill Response

UNIVERSITY OF MAINE

RAYMOND H. FOGLER LIBRARY

A Practical Guide to Chemical Spill Response

John Hosty

with

Patricia E. Foster

VNR VAN NOSTRAND REINHOLD
      New York

Copyright © 1992 by Van Nostrand Reinhold
Chapter 2 copyright © 1992 Gordon Norman

Library of Congress Catalog Card Number 91-47626
ISBN 0-442-00569-5

All rights reserved. No part of this work covered by the copyright hereon may be reproduced or used in any form by any means—graphic, electronic, or mechanical, including photocopying, recording, taping, or information storage and retrieval systems—without written permission of the publisher.

Printed in the United States of America

Van Nostrand Reinhold
115 Fifth Avenue
New York, New York 10003

Chapman and Hall
2-6 Boundary Row
London, SE1 8HN, England

Thomas Nelson Australia
102 Dodds Street
South Melbourne 3205
Victoria, Australia

Nelson Canada
1120 Birchmount Road
Scarborough, Ontario M1K 5G4, Canada

16 15 14 13 12 11 10 9 8 7 6 5 4 3 2 1

Library of Congress Cataloging-in-Publication Data
Hosty, John.
   A practical guide to chemical spill response/John Hosty.
    p. cm.
   Includes bibliographical references and index.
   ISBN 0-442-00569-5
   1. Chemical spills—Safety measures. 2. Hazardous substances—Safety measures. I. Title.
TD196.C45H67 1992
628.9—dc20   91-47626
                                                                              CIP

To Lauren, Drew and Sean

Contents

Preface ix

1. Pre-Incident Activities   **1**

   1. Contingency Planning   1
   2. Training   14
   3. Review of Government and Industry Organizations   16

2. Crisis News Management for Hazardous Materials Incident Responders and Incident Managers   **25**

   Gordon Norman
   1. The News Media as Entertainers   25
   2. The News Media as Environmental Opinion Makers   26
   3. The Growing Public Discontent with Modern Business Ethics   27
   4. What to Expect when an Incident Occurs   28
   5. The "On Camera" Interview   33
   6. A Note about Media Kits   35

3. Personal Protection Equipment   **37**

   1. Respiratory Protection   37
   2. Protective Clothing   48
   3. Decontamination Procedures   60

4. Incident Risks and Safety   **68**

   1. Flammability of Liquids   68
   2. Flammability of Gases   70
   3. Corrosivity   82
   4. Reactivity   83
   5. Toxic Risk   84
   6. Oxygen Deficiency/Enrichment   91

	7. Hazard Indication	92
	8. Other Incident Risks	103
	9. Incident Assessment, Approach, and Monitoring	105
5.	**Spill Control**	**115**
	1. Containment	115
	2. Recovery	131
	3. Drums and Cylinders	140
6.	**Air Monitoring**	**148**
	1. Values Monitored	149
	2. Safety in Air Monitoring	151
	3. Air Monitoring Instruments	151
7.	**Tank Trucks**	**158**
	1. Safety Design Features	158
	2. Cargo Tank Types	159
	3. Valves	167
8.	**Rail Cars**	**171**
	Joe P. Riddle	
	1. General Description	171
	2. Response to Tank Car Incidents	174
	3. Types of Tank Cars	176
9.	**Intermodal Tank Containers**	**179**
	Joe P. Riddle	
	1. Description of Containers	179
	2. Tank Container Fittings	180
	3. Tank Container Markings	182
	4. Responses to Intermodal Tank Container Incidents	183

Glossary — **185**

Index — **189**

Preface

Our interest in the many facets of spill response began over ten years ago in John's position as a ship's Master in oil tankers and has grown through the years with his work both as an educator and within the environmental services industry. One thing that always seemed to be missing in both areas was a "how to" guide-book that would serve as both a simple day to day resource and as a method of orientation to the field of spill response. In writing *A Practical Guide to Chemical Spill Response* we kept this requirement at the front of our minds.

Safe spill response is a complex, albeit uncomplicated, operation that requires the integration of numerous techniques and technologies. Governmental and industrial agencies are constantly changing and trying to expand their influence on the field. Safe and effective response pertains largely to issues such as health and safety of personnel at incident sites along with the protection of nearby populations. This can only be achieved with careful pre-planning and assessment of the potential for incidents. This book should be an effective reference during such planning stages.

There are people from many walks of life who may have to respond to spills of hazardous materials. The simplest situation may be a person on a shop floor who has to clean up a small spill of a relatively innocuous product; the most extreme situation may be a group of industrial and municipal responders dealing with a large incident involving railcars, tank trucks and a multitude of high hazard materials. No matter how complex or simple the situation, this book will ensure that a safe, effective response is achieved under all circumstances and conditions.

A Practical Guide to Chemical Spill Response reviews all phases of spill response, from planning to final recovery and restoration of a spill site and assumes that the reader has a rudimentary understanding of the physical and chemical properties of hazardous materials. It looks in detail at how to select respiratory protective equipment, chemical protective clothing, and air monitoring instruments, and how to use such items in practical field settings. It shows by practical guide and illustration the "how to" of spill response to spills involving oils, chemicals, and other hazardous materials. It provides a self study grounding in all aspects and reflects the requirements of current legislative trends for training in the topic.

tion and Recovery Act (RCRA), Title 44 of the Code of Federal Regulations, and certain Occupational Safety and Health Administration (OSHA) regulations. In Canada, the only direct requirements for development of a contingency plan are, federally, under the Transportation of Dangerous Goods Act for certain dangerous goods, and provincially, under certain Ministry of Environment legislative documents.

An essential part of contingency plan development is reviewing response capabilities and equipment availability, both internally and externally. Any shortfalls recognized should be remedied as part of the contingency planning process. A contingency plan should include the elements described below.

1A. Contact List

The contact list should include the names and telephone numbers of all organizations and personnel, both internal and external, who may be involved in response to an incident that the plan is written for. During assembly of the plan, all numbers should be confirmed by telephoning and ensuring that the designated individual is available at the assigned telephone number. Where possible, 24-hour contact numbers should be given. Once assembled, this list should be audited every 90 days.

1B. Regulatory Review

Many actions undertaken during a spill are driven by regulatory requirements, particularly those of alerting and reporting; indeed, legislation in many jurisdictions requires the assembly of a contingency plan.

The regulatory review section of the plan should highlight the legislation that has an impact on spill response and management. While no two jurisdictions are alike, there may be federal, national, provincial, state, or local legislation that affects the following:

- Transportation of hazardous materials
- Waste registration, transportation, and disposal
- Spill reporting
- Spills on marine vessels, into navigable waters
- Registration of hazardous materials
- Worker awareness
- Responder training

These documents should be reviewed frequently as their content may change.

Preface

Our interest in the many facets of spill response began over ten years ago in John's position as a ship's Master in oil tankers and has grown through the years with his work both as an educator and within the environmental services industry. One thing that always seemed to be missing in both areas was a "how to" guide-book that would serve as both a simple day to day resource and as a method of orientation to the field of spill response. In writing *A Practical Guide to Chemical Spill Response* we kept this requirement at the front of our minds.

Safe spill response is a complex, albeit uncomplicated, operation that requires the integration of numerous techniques and technologies. Governmental and industrial agencies are constantly changing and trying to expand their influence on the field. Safe and effective response pertains largely to issues such as health and safety of personnel at incident sites along with the protection of nearby populations. This can only be achieved with careful pre-planning and assessment of the potential for incidents. This book should be an effective reference during such planning stages.

There are people from many walks of life who may have to respond to spills of hazardous materials. The simplest situation may be a person on a shop floor who has to clean up a small spill of a relatively innocuous product; the most extreme situation may be a group of industrial and municipal responders dealing with a large incident involving railcars, tank trucks and a multitude of high hazard materials. No matter how complex or simple the situation, this book will ensure that a safe, effective response is achieved under all circumstances and conditions.

A Practical Guide to Chemical Spill Response reviews all phases of spill response, from planning to final recovery and restoration of a spill site and assumes that the reader has a rudimentary understanding of the physical and chemical properties of hazardous materials. It looks in detail at how to select respiratory protective equipment, chemical protective clothing, and air monitoring instruments, and how to use such items in practical field settings. It shows by practical guide and illustration the "how to" of spill response to spills involving oils, chemicals, and other hazardous materials. It provides a self study grounding in all aspects and reflects the requirements of current legislative trends for training in the topic.

1
Pre-Incident Activities

This book is intended to be a concise reference for those who have to respond to spills and other incidents involving hazardous materials. The majority of incidents, whether they occur during transportation, in an industrial plant, or elsewhere, will be small and readily accessible. We do not deal with large-scale incidents such as those that occurred at Mississauga, Bhopal, or Crescent City, every day. Regardless of their size, however, all incidents are easier to resolve if they have been assessed in the planning stages. Injuries and chemical exposures to response personnel and others will be reduced if trained people respond in a safe manner with adequate supplies of the correct equipment.

This first chapter reviews steps to be taken prior to the occurrence of an incident, including

- Development of a contingency plan
- Training of personnel
- Review and assessment of government, industrial, and other resources

1. CONTINGENCY PLANNING

A contingency plan is best described as an assessment of potential incidents and the development of predetermined sequences of events to deal with them. A good contingency plan should be readable, understandable, and easy to follow. In an emergency situation it is a useful tool for responders in planning and implementing their management of an incident.

Contingency plans are required under several jurisdictions and documents, notably, in the United States, the Superfund Amendments and Reauthorization Act, Title III (SARA Title III), the Resource Conservation and Recovery Act (RCRA), Title 44 of the Code of Federal Regulations, and certain Occupational Safety and Health Administration (OSHA) regulations. In Canada, the only direct requirements for development of a contingency plan are, federally, under the Transportation of Dangerous Goods Act for certain dangerous goods, and provincially, under certain Ministry of Environment legislative documents.

An essential part of contingency plan development is reviewing response capabilities and equipment availability, both internally and externally. Any shortfalls recognized should be remedied as part of the contingency planning process. A contingency plan should include the elements described below.

1A. Contact List

The contact list should include the names and telephone numbers of all organizations and personnel, both internal and external, who may be involved in response to an incident that the plan is written for. During assembly of the plan, all numbers should be confirmed by telephoning and ensuring that the designated individual is available at the assigned telephone number. Where possible, 24-hour contact numbers should be given. Once assembled, this list should be audited every 90 days.

1B. Regulatory Review

Many actions undertaken during a spill are driven by regulatory requirements, particularly those of alerting and reporting; indeed, legislation in many jurisdictions requires the assembly of a contingency plan.

The regulatory review section of the plan should highlight the legislation that has an impact on spill response and management. While no two jurisdictions are alike, there may be federal, national, provincial, state, or local legislation that affects the following:

- Transportation of hazardous materials
- Waste registration, transportation, and disposal
- Spill reporting
- Spills on marine vessels, into navigable waters
- Registration of hazardous materials
- Worker awareness
- Responder training

These documents should be reviewed frequently as their content may change.

1C. External Organizations

Within a reasonable distance of perceived incident sites there may be many organizations that can be called to assist in response. This is particularly pertinent for off-site transportation spills, where the distances involved may make it difficult to bring internal resources to bear within a reasonable time frame. External bodies that should be detailed include:

- CANUTEC/CHEMTREC
- TEAP (provided a contract is in place)/CHEMNET
- Local spill co-ops
- State/provincial chemical association groups
- Product mutual-aid groups (e.g., CHLOREP for chlorine spills)
- Environmental Protection Agency/Ministries (federal and state/provincial)
- Coast Guard
- Fire departments
- Police departments
- Other municipal organizations and facilities (e.g., water treatment plants and spill control facilities)
- Paramedics
- Environmental contractors

Further details on the functions of these organizations appear later in this chapter.

1D. Command Structures

Many parties may be involved in the resolution of incidents—for example, government agencies, industry organizations, municipal agencies, product owners, and transportation operators. So that these parties may be readily integrated into incident activities, it is important that a clear incident command system (ICS) be pre-established and included in the contingency plan. OSHA specifically requires the use of an ICS for responses in the United States (29 CFR 1910.120q). The topic of incident command is comprehensively addressed in several other documents and publications. A recommended reference is the *Hazardous Materials Response Handbook*, published by the National Fire Protection Association.

1E. General Information

The general information section should include:

- A *vicinity plan*, showing the general position of the facility in relation

to other local industrial facilities, mutual-aid organizations, municipal responders, etc. (See Figure 1.1.)
- A *general arrangement plan,* showing the layout of the site or sites at which the plan is designed to be used. Particular attention should be paid to the location of transfer points, storage systems, and pipelines. (See Figure 1.2.)
- A *transportation route plan,* showing routes along which hazardous materials will be shipped, potential exposures, mutual-aid centers, equipment locations, etc. (See Figure 1.3.)
- An *emergency equipment plan,* a diagram showing the position of items such as fire-fighting, spill management, respiratory protective, personal protective, and other such equipment. Additionally, where provided, emergency shutdown locations and the zones controlled, water drafting points, hydrants, electrical shutoffs, and telephone locations should be shown. (See Figure 1.4.)
- The *use of the contingency plan,* detailing the circumstances for which the plan has been written and under which the plan should be adhered to. This section should also describe how the plan is to be implemented and by whom, along with a list of events for which the plan may be implemented.
- *Copies* of the contingency plan: a list of copies that are held, where, and who is responsible for their update. This is important so that plan amendments can be circulated to all copy holders.
- The *process for amendment*—generally, the plan should be amended whenever changes occur (i.e., telephone numbers, new equipment purchased, new products handled, old materials and information to be expunged). The plan should be reviewed and exercised at least annually and amendments made to reflect any findings.
- The *evacuation plan,* including details of who can order an evacuation and the logistics of its accomplishment. Many jurisdictions designate specific government agencies and/or representatives as the authority to order an evacuation. The contingency plan should review evacuation routes and muster points for a variety of circumstances and conditions. These include variations in wind and temperature, potential plume development, precipitation, type of product assessed, and its geographic position within the facility. Typical factors of routing include protection from fire, explosion, and vapor exposure, availability of suitable muster points and adequate shelter.

1F. Follow-up Activities

Once the initial phases of an incident have been addressed, the plan should continue to deal with activities that ensue, including the following.

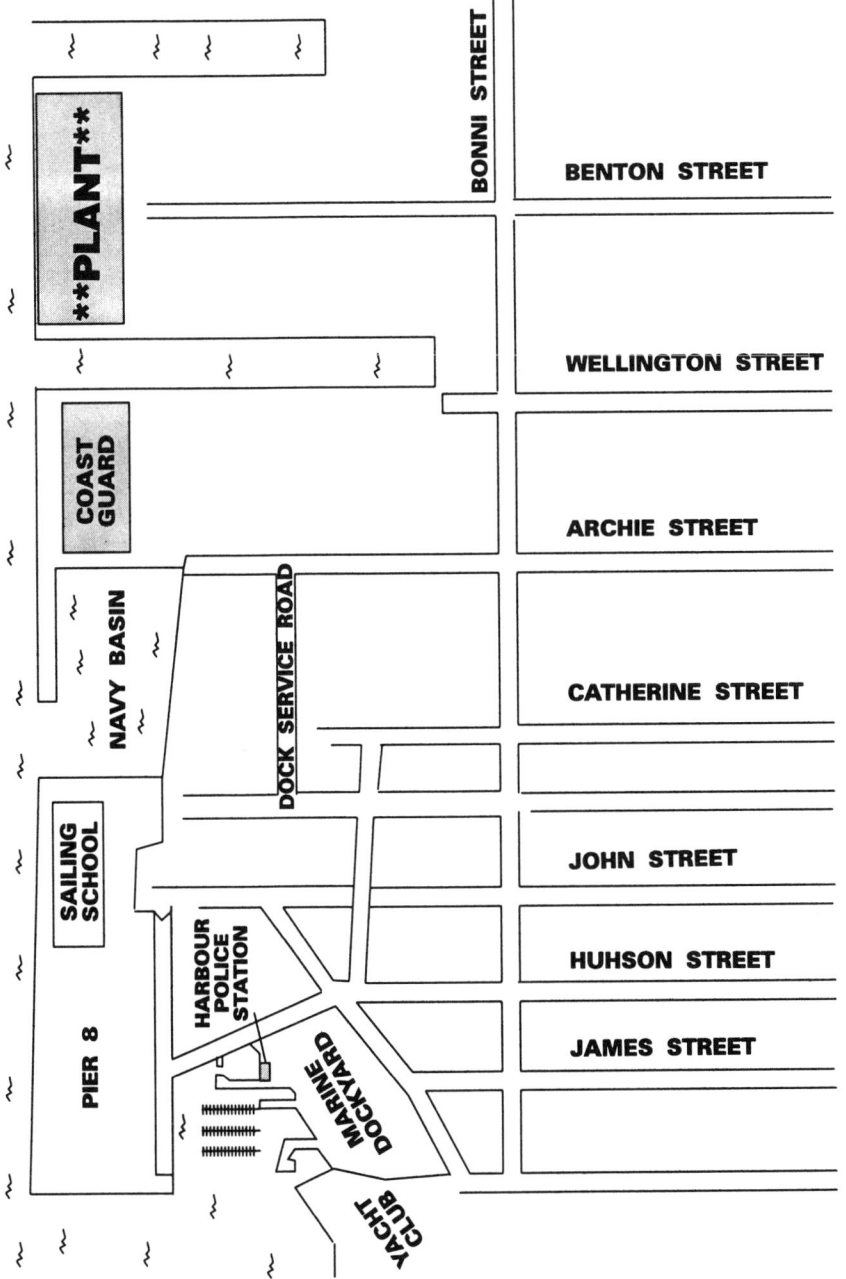

FIGURE 1.1. Vicinity Plan

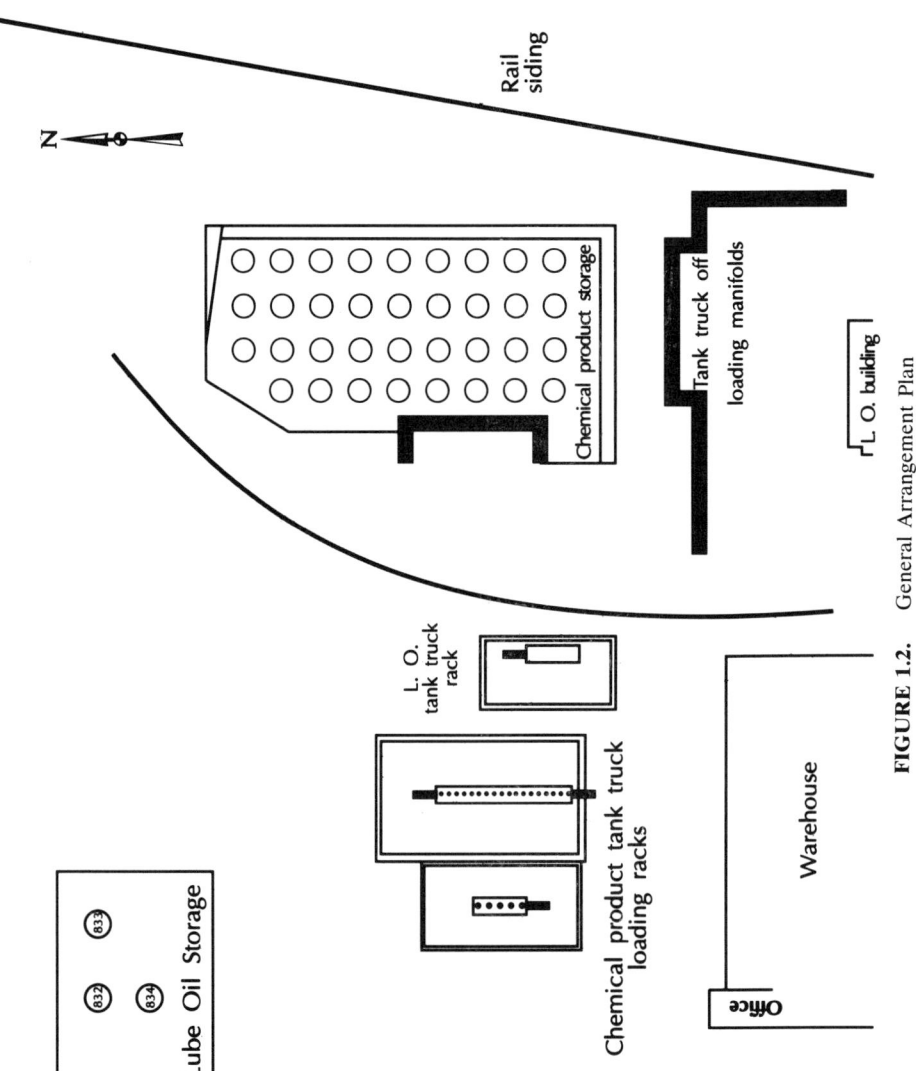

FIGURE 1.2. General Arrangement Plan

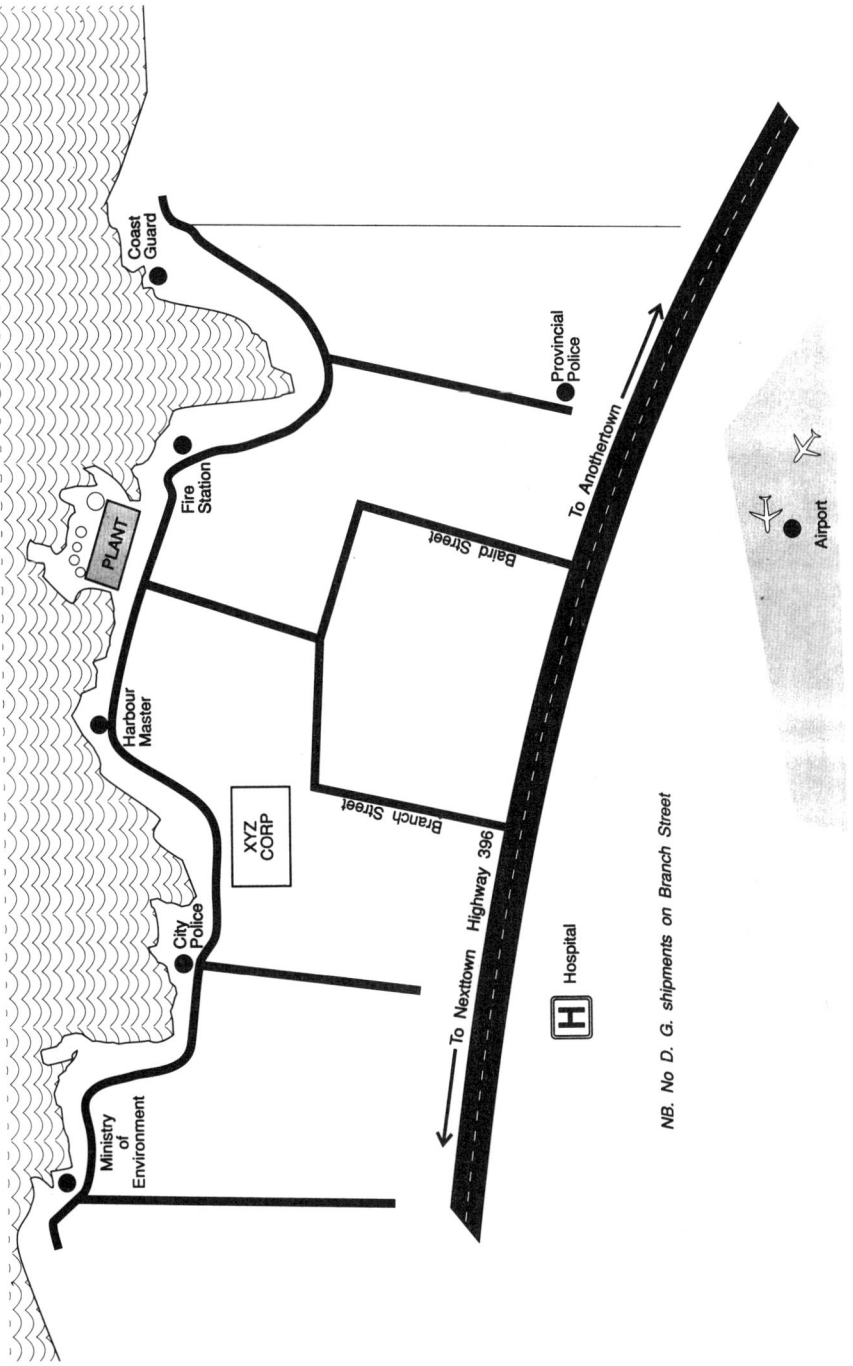

FIGURE 1.3. Transportation Route Plan

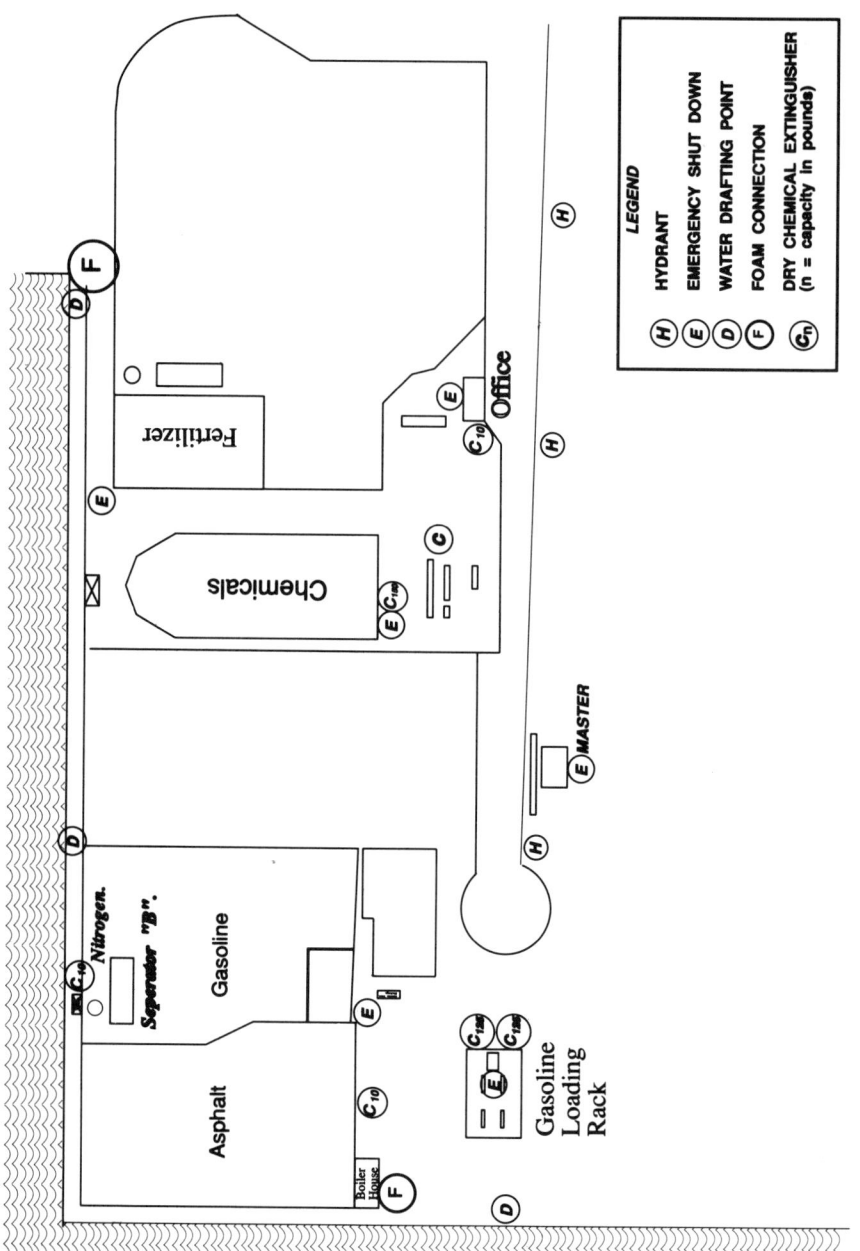

FIGURE 1.4. Emergency Equipment Plan

- Storage of contaminants, including details of how, where, and for how long recovered materials can be stored. Most jurisdictions place a time limit on the storage of recovered materials.
- Registration procedures, including details of where contaminants and recovered product/debris can be analyzed, by certified laboratories, and how they are to be registered with environmental agencies.
- Removal and disposal, which may be effected only by duly licensed organizations. Contractual arrangements should be in place with licensed waste haulers, and disposal sites should be predesignated. Of particular importance is a list of facilities that are not to be accessed for disposal; it is prudent to audit all designated facilities while the plan is being developed.
- Decontamination, since all equipment used in handling an incident must be properly decontaminated and passed as fit for reuse prior to final storage. Decontamination procedures will depend on the product being handled, but may include solvent washing, detergent washing, rinsing, drying, and finally wipe testing. All equipment that cannot be properly cleaned should be disposed of as contaminated material. Damaged equipment should be decontaminated prior to being disposed of.
- Reporting. In addition to legislated immediate reporting requirements and follow-up reports, some internal reporting procedures should be implemented. All initial reporting is included as part of the response forms, but internal follow-up is best achieved by use of forms similar to those depicted in Figures 1.5, 1.6, and 1.7.

1G. Public Relations

Invariably, the incidents alluded to by a plan will attract the attention of both the public and the media. Frequently, failure to communicate information to the media results in the appearance of incorrect and/or embarrassing accounts. For this reason, great care should be taken to communicate the facts of an incident, its causes, and the issues that surround it.

One or more people should be predesignated by the plan to address the public and the media. No other personnel should be involved in this interface. Chapter 2 discusses in detail issues relating to public and media relations.

1H. Incident Management Techniques

The meat of any contingency plan is a review and detail of the management of spills involving all products, both within a facility and where required off-site. Incident management techniques should address containment

Environmental Incident Report

1. Plant: _____

2. Source of Spill: _____

3. Time: _____ Date: _____ Period: _____

4. Spill reported by: _____ Spill Site Coordinator _____

5. Material and Estimated quantity spilled: _____

6. Cause of Spill: _____

7. Medium affected: Water ____   Air ____   Soil ____   Other ____

8. Area affected: _____

9. Personnel notified.   Name: _____   Date: _____   Time: ____

10. Government Agencies & Others Notified: Name   Date   Time

	Name	Date	Time
Ministry of Environment			
Fuels Safety			
Regional Municipality			
T D G			
Coast Guard			
Product Owner			
Transportation Owner			

11. Damage to Property & Environment _____

12. Incident under control: Yes ____   No ____

13. Containment and clean-up methods

    Internal Methods   Contractor
    _____   _____
    _____   _____
    _____   _____

14. Estimated cost of spill: $ _____   Report completed by: _____

                                         Date: _____

FIGURE 1.5.   Environmental Incident Report

Emergency Response Questionaire

Date:

Time call received: _____

Name of caller: _____

Organization: _____

Callers phone #: _____

Consignee name: _____

Site contact: _____

Phone #: _____

Incident address: _____

Town: _____

Time of incident: _____

Name of product: _____

Class of product: _____

Quantity in shipment: _____

Quantity spilled: _____

Physical state: Liquid ___ Solid ___ Gas ___

Incident Cause:

- Accident ☐
- Leak ☐
- Fire ☐
- Mechanical failure ☐
- Other _____

Describe incident:
_____
_____
_____
_____
_____

Who was responsible for product?
_____
_____
_____
_____
_____

Geography:

Population proximity: _____

Watercourse proximity: _____

Water supply proximity: _____

Industry proximity: _____

Drains & sewers: _____

Owner:

Road vehicles: _____

Rail vehicles: _____

Marine vessel: _____

Name of vessel: _____

Weather:
Fair ____ Windy ____ Raining ____
Overcast ____ Foggy ____ Temp ____

FIGURE 1.6. Emergency Response Questionnaire

Emergency Response Questionaire (cntd)

Who advised?

Police _____

Fire Dept _____

CANUTEC _____

M.O.E. _____

Transport Canada _____

Regional Municipality _____

Others _____

Additional remarks:

_____
_____
_____
_____
_____
_____
_____

Who is at scene?

Police _____

Fire Dept _____

M.O.E. _____

Transport Canada _____

Regional Municipality _____

Others _____

Questionaire completed by:

Name: _____

Title: _____

Date: _____

CONFIDENTIAL

FIGURE 1.7. Emergency Response Questionnaire (contd.)

techniques, reporting, recovery, and disposal methods. Figure 1.8 offers an example.

1I. Equipment Review

A major part of the contingency planning process is to review available equipment and its applicability to perceived responses. Every situation will be different. Table 1.1 is a suggested list.

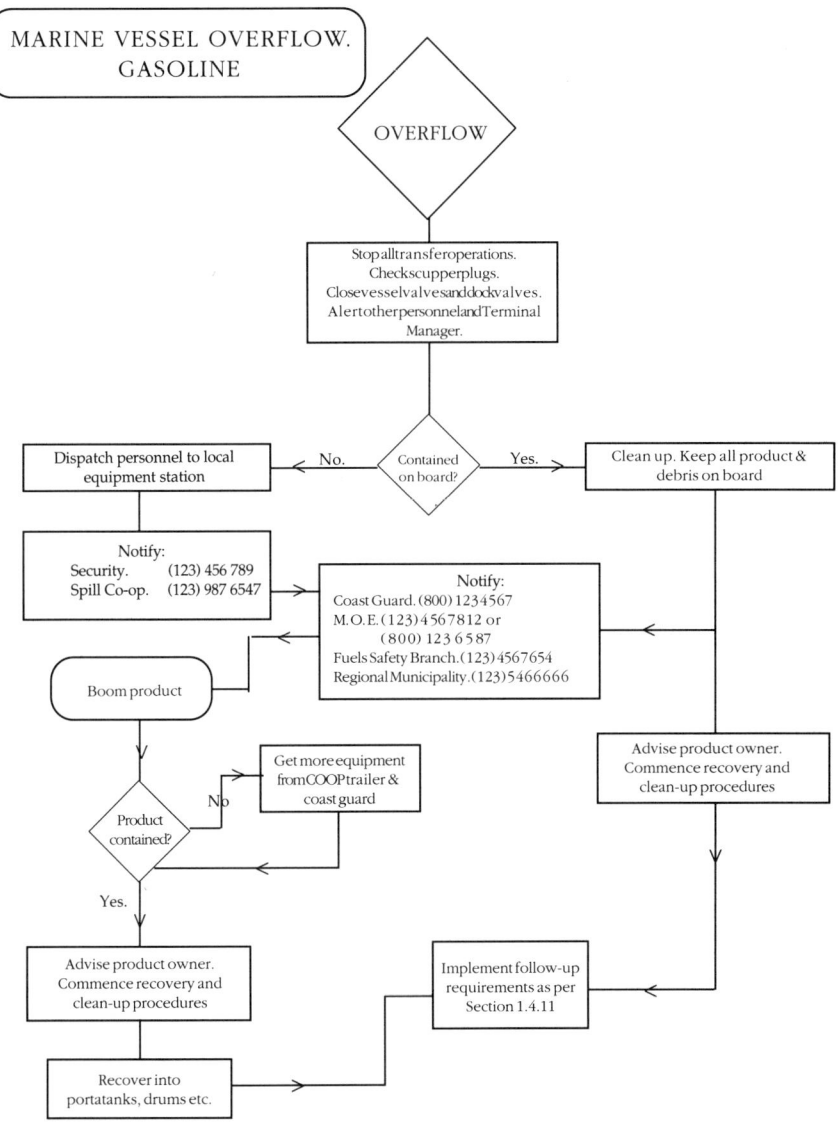

FIGURE 1.8. Action Summary Form

TABLE 1.1 Suggested Equipment List

Containment/Deflection	**Miscellaneous**
Boom, polyethylene, 6″ diameter	Analyzer, colorimetric tube
Sheets, plywood	Analyzer, combustible
Snow fence	Analyzer, oxygen
Sheeting, polyethylene	Analyzer, P.I.D.
	Axe
	Brooms
	Catchbasin covers
Electrical	Caution tape
	Chemical protective clothing (as required)
Cellular phone	Clamps, pipe, assorted
Continuity tester	Crowbar
Electrical cabling	First Aid Kit
Extension cords, 30 metre	Hacksaw 8″
Flashlights	Hacksaw 14″
Generator 1 kW to 5 kW	Hard hats
Generator 5 kW to 10 kW	Hose fittings
Grounding cables, 30 meters	Hose, compressed air
Grounding clamps	Hose, discharge
Grounding rods, 2 meters	Hose, suction
Lights, flood with stands	Mops
Portable radios	Picks
	Pitchforks
	Rakes
	Recovery drums
	Respiratory protective equipment (as required)
Sorbents and Neutralizers	Sand
	Sealing compounds
Absorbent pads	Shovels
Absorbent rolls	Sledgehammer large
Absorbent sheets	Sledgehammer small
Adsorbents, chemical	Tape, duct, 1 1/2″
Chicken wire	Tape, electrical
Neutralizers, chemical, solid	Wrench, adjustable 6″
Neutralizers, liquid	Wrench, adjustable 10″
Sand	Wrench, adjustable 12″
Spill socks	Wrench, spark plug
Straw	Wringer on stand

2. TRAINING

Prior to becoming involved in responding to hazardous materials spills and incidents, personnel must be trained in all the aspects of incident resolution. The primary goal of training in the field of hazardous materials emergency response is safety and worker protection. Several statutes in the United States and Canada provide specific training requirements. These are categorized as follows:

- Right to know
- Transportation
- Spill response

2A. Right to Know

Every person has a legislated right to be made aware of the dangers to which he or she will be exposed when dealing with hazardous materials in the workplace. For the purpose of "right to know," the definition of hazardous materials is as construed by the Workplace Hazardous Materials Information System (WHMIS) in Canada and the Hazard Communication Standard (HAZCOM, 29 CFR 1910.1200) in the United States. These documents are very similar in nature and detail how worker awareness will be achieved—this being essentially by familiarization with the products handled and by use of material safety data sheets (MSDS). There is currently some debate as to what constitutes a workplace, however, so it is prudent to assume that any setting where a worker may be exposed to hazardous materials, which could include a spill site, will be affected by this type of legislation.

2B. Transportation

The transportation of hazardous materials is addressed by Title 49 CFR in the United States and by the Transportation of Dangerous Goods Act and Regulations in Canada. The scope of substances addressed under these documents is, generally speaking, not as broad as those detailed in right-to-know legislation. The level of hazard is often higher.

Part 9 of the Transportation of Dangerous Goods Regulations requires that all who handle, offer for transport, or transport dangerous goods be trained in, among other topics, emergency action, handling, and safety requirements.

Other documents that have an impact include the International Maritime Dangerous Goods Code (IMDG), the Regulations for the Transportation Dangerous Goods by Rail (Canada), and the International Civil Aviation Organization's Technical Instructions for the Transportation of Dangerous Goods by Air.

2C. Spill Response

There is a marked difference in approach to spill response training between Canada and the United States. Through the Superfund Amendments and Reauthorization Act (SARA), President Ronald Reagan directed the Occupational Safety and Health Administration (OSHA) to develop and institute standards for workers at Superfund sites and certain Resource Recovery and Conservation Act (RCRA) sites. This directive also included the setting of another, slightly less onerous, standard for those workers attending emergencies involving hazardous materials. Such persons must receive a minimum of 24 hours training annually. These standards are promulgated in Title 29 CFR, 1910.120.

In defining the content of such training, OSHA specifically references the National Fire Protection Association's (NFPA) Standard for Professional Competence of Responders to Hazardous Materials Incidents (NFPA 472). This standard recognizes four levels of competency: first responder awareness, first responder operational, hazardous materials technician, and hazardous materials specialist. This standard is currently being reviewed by government, industrial agencies, and associations in both the United States and Canada.

3. REVIEW OF GOVERNMENT AND INDUSTRY ORGANIZATIONS

Several bodies in the United States and Canada have been formed for the purpose of mutual aid in emergency and spill situations.

3A. Canada

The organizations in Figure 1.9 contribute to mutual aid in Canada.

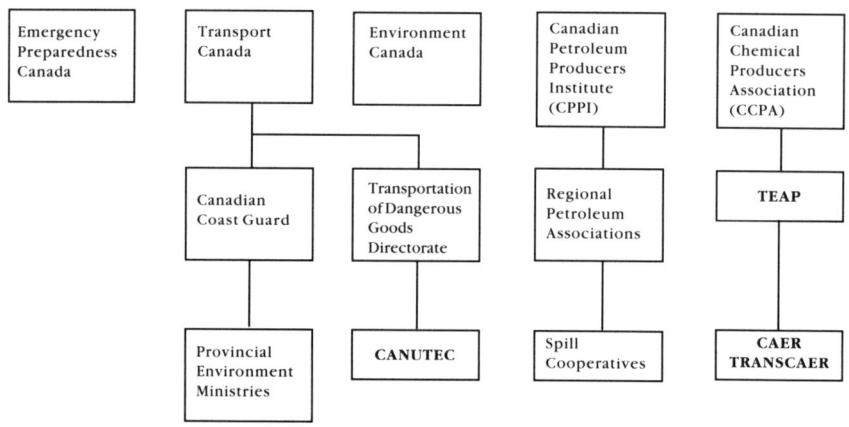

FIGURE 1.9. Principal Government and Industry Emergency Response Infrastructures in Canada

3. Review of Government and Industry Organizations 17

a. Transport Canada

Transport Canada operates several initiatives through its various directorates, including the Transport Dangerous Goods directorate and the Canadian Coast Guard.

Transport Dangerous Goods operates the Canadian Transportation Emergency Centre (CANUTEC). CANUTEC is a 24-hour service that is available without charge to the caller from anywhere in Canada by calling (613) 996-6666. In an incident, CANUTEC will put response personnel in touch with the consignor and provide product information.

In recent years CANUTEC has become more aggressive in its guidance to incident responders. The centre is staffed by a highly skilled, competent team, all of whom have significant training and experience in the field of emergency response.

Additional functions of CANUTEC include acting as a consignor's 24-hour contact and publication of the *Dangerous Goods Guide to Initial Emergency Response* (rev. 1986).

All spills involving marine vessels, or those that enter navigable waterways, are reportable to the Canadian Coast Guard at (800) 265-0237. Additionally, the Coast Guard maintains stocks of recovery equipment at various points across Canada. Of particular note is their maintenance of some large-scale recovery equipment.

b. Environment Canada

Environment Canada is concerned primarily with the protection of federal lands (i.e., lands administered by the Department of National Defense, railways, Crown properties, etc.). The department has emergency branches across the country; in spill situations that fall within their jurisdiction, a representative will attend.

c. Provincial Environment Agencies

Every province in Canada attends to its own environmental affairs. The functions of these agencies include waste management, spill management, and spill reporting. The approaches vary from province to province, but the fundamental methodology is the same. All spills are reportable to the appropriate environment ministry unless specifically exempt.

d. National Transportation Agency

Among other functions, the National Transportation Agency (NTA) has assumed the spill response duties of what was the Canadian Transport Commission (CTC). While the NTA will render advice upon request, their primary function in this field is to investigate rail incidents. All rail hazardous materials incidents are reportable to the NTA at (819) 997-0344.

e. Canadian Chemical Producers Association

The Canadian Chemical Producers Association (CCPA) is the organizing body for the Transportation Emergency Assistance Plan (TEAP). This network of response centers allows for mutual aid between member companies. In order to utilize TEAP and its equipment, a company must be both a CCPA member and a signatory to the TEAP contract. At the time of writing, only CCPA members can become signatories to TEAP.

In addition to equipment and response personnel, TEAP requires members to make available technical advisors at an incident scene. Figure 1.10 illustrates the location of TEAP Regional Response Centres (RRCs).

f. Canadian Petroleum Products Institute

The Canadian Petroleum Products Institute (CPPI) is somewhat akin to the CCPA in its functions. Through its associated research division, the Petroleum Association for the Conservation of the Canadian Environment (PACE), CPPI provides equipment research and development in the field of spills. Additionally, regional petroleum associations across Canada—namely, the British Columbia, Prairie, Ontario, Quebec, and Atlantic Petroleum Associations—which maintain both mobile and fixed facilities in many places across Canada. Each facility is equipped with suitable equipment for perceived potential spill problems in the area.

3B. United States

The organizations in Figure 1.11 contribute to mutual aid in the United States.

a. Federal Emergency Management Agency

The Federal Emergency Management Agency (FEMA) provides training in emergency response, not only for hazardous materials incidents but also for other civic emergencies and disasters (such as tornados, earthquakes, and floods). Additionally, this agency produces many references and guidelines for planning and plan development.

b. U.S. Environmental Protection Agency

The U.S. Environmental Protection Agency (USEPA) addresses any and all issues that affect the environment. While spills are handled by state EPAs, the USEPA may be called in to assist in larger-scale incidents. Any events that occur on federal property or that cross state boundaries come under the jurisdiction of the USEPA.

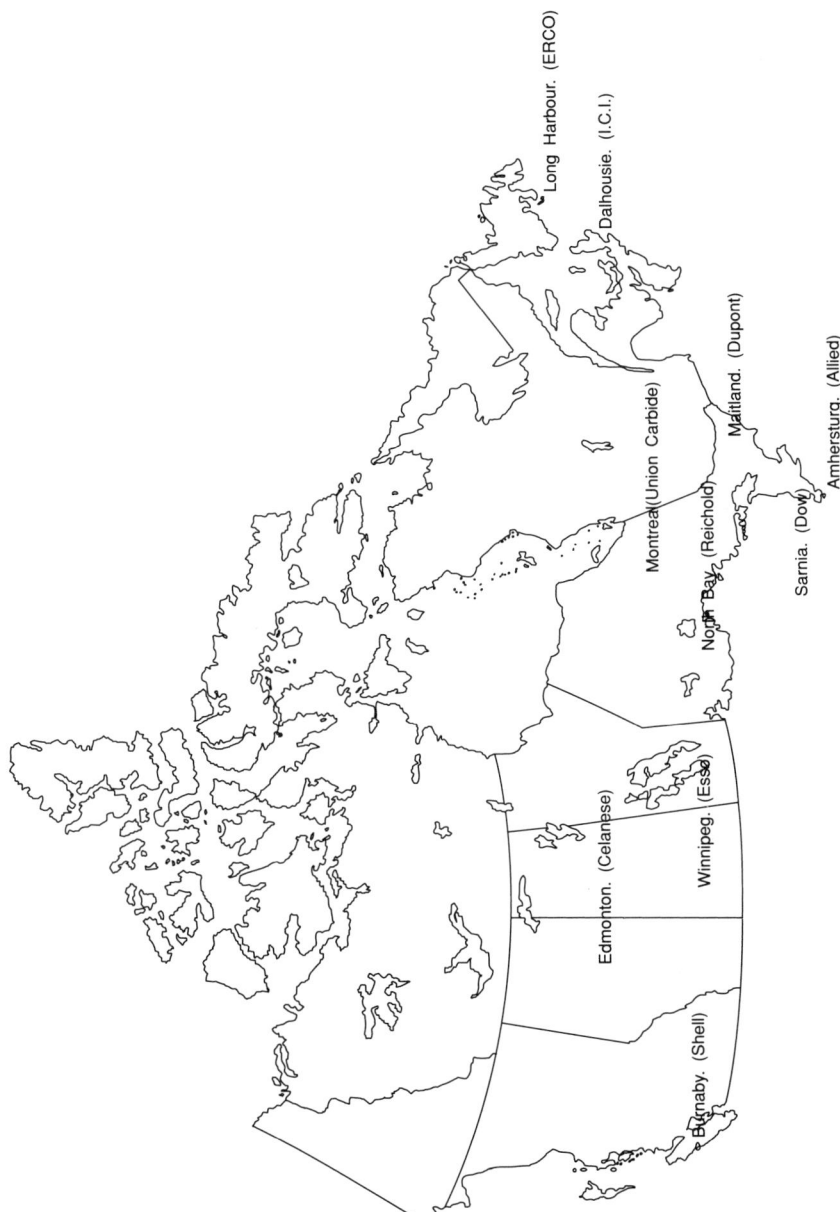

FIGURE 1.10. TEAP Regional Response Centres

20   Pre-Incident Activities

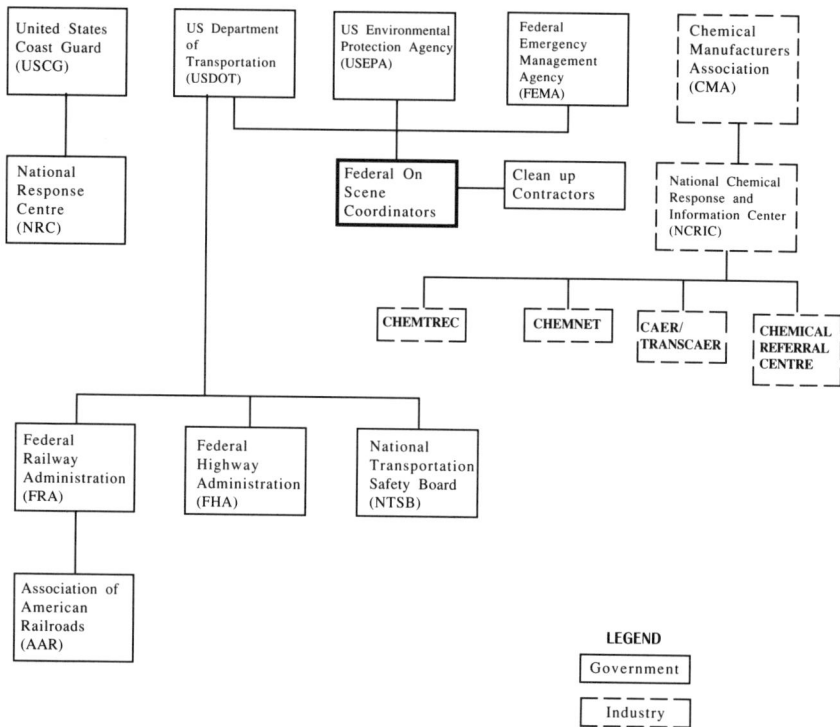

FIGURE 1.11. Principal Government and Industry Emergency Response Infrastructures in the United States

c. U.S. Coast Guard

All spills into navigable waters and all spills involving vessels are reportable to the U.S. Coast Guard (USCG). The USCG is somewhat more paramilitary in its organization than its Canadian counterpart and assumes far greater responsibilities and action in a spill situation.

In addition to the reporting requirement, the Coast Guard operates the National Response Center (NRC). A telephone call logged at the NRC satisfies federal reporting requirements and alerts on-scene coordinators to environmental incidents in their geographic jurisdictions.

d. Chemical Manufacturers Association

Through the National Chemical Response Information Centre (NCRIC), the Chemical Manufacturers Association (CMA) maintains several initiatives, including:

3. Review of Government and Industry Organizations   21

- CHEMTREC
- CHEMNET
- Chemical Referral Center
- Hazardous Materials AV Lending Library
- Emergency response training
- CAER/TRANSCAER

CHEMTREC
CHEMTREC, (800) 424-9300, provides 24-hour communications with emergency response personnel, chemists, product specialists, and the medical departments of more than 6000 hazardous materials shippers and carriers. Additionally, CHEMTREC coordinates the activities of several mutual-aid groups whose teams are specially trained to deal with emergencies involving:

- Chlorine
- Phosphorus
- Hydrogen cyanide
- Hydrogen fluoride
- Compressed gases
- Liquefied petroleum gases
- Swimming pool chemicals
- Pesticides (North Carolina Emergency Action Team)
- Liquid fluorine and nitrogen tetroxide (DoD)

A call to CHEMTREC will, if warranted, initiate a response from one of these groups.

CHEMTREC maintains an exhaustive database containing over 570,000 material safety data sheets, provided by product manufacturers, in an on-line computer imaging system. Sheets are provided directly to emergency service organizations via fax during emergencies.

Additional CHEMTREC services include:

- Direct access to information from the major railroads
- Foreign language interpretation
- Access to a repository of specialized containers for isolating spilled products

CHEMTREC in the United States and CANUTEC in Canada are organizations with similar functions.

CHEMNET
CHEMNET (Chemical Network), the chemical industry's emergency response mutual-aid network, was established in 1985 to expand the emergency response capabilities of the chemical industry, to provide timely

emergency response and technical assistance at the scene of serious hazardous materials transportation incidents in minimum time, and to meet the growing expectations and safety concerns of state and local governments and citizens.

CHEMNET is a mutual-aid emergency response system comprised of emergency response teams from chemical companies and private for-hire emergency response/spill cleanup contractors. When activated by participating companies in the event of a serious hazardous materials transportation emergency, CHEMNET allows members to share their expertise and provide "experts" on-site in a timely manner to assist local authorities in managing and mitigating the emergency in a safe and timely manner.

By establishing a vast network of industry emergency response teams, CMA is helping many communities deal with their concerns about chemical incidents. CHEMNET members can provide on-site technical expertise and specialized equipment for many small and medium-sized communities that do not have these capabilities.

The CHEMNET program includes approximately 250 emergency response teams, and the program continues to expand. The CHEMNET program is part of an overall initiative to improve emergency response to serious chemical transportation incidents in minimum time.

CHEMNET in the United States and TEAP in Canada are organizations with similar functions.

Operation/Activation CHEMNET operates through CHEMTREC, located at CMA headquarters in Washington, D.C. CHEMTREC maintains current listings of members, emergency response team locations and capabilities, and is operational 24 hours a day, 7 days a week.

The CHEMNET network is activated when a participating chemical shipper is notified that a serious chemical distribution incident (SCDI) involving one of its chemicals has occurred and expert assistance is needed at the scene. An incident can be classified as an SCDI if it meets any of the following criteria:

- An injury or death results from exposure to the chemical(s) involved.
- The chemical(s) involved is very toxic, unstable, or flammable.
- A large quantity of the chemical(s) is released or is in a container that might rupture.
- The size/rate of the spill or leak is large, or the potential for environmental damage is great.
- Field transfer of the product(s) will be required.
- Assistance is requested by the caller.
- The presence of an industry team might prevent mishandling or overreaction.

- A large evacuation has been ordered.
- A derailment involves hazardous materials and the tank car is not upright.
- A tank truck containing hazardous materials has been involved in a rollover accident.

For purposes of activating responders, the shipper determines whether an SCDI exists and whether the incident merits activation of CHEMNET. The shipper may also declare any incident an SCDI, even though the initial information received via CHEMTREC does not meet one of the SCDI criteria.

If the shipper cannot be contacted, CHEMTREC can make the determination to activate CHEMNET. If the shipper is unable to respond in a timely manner because of lack of availability or proximity, CHEMTREC provides the shipper with the name and location of the closest emergency response team they may contact for mutual aid. Through CHEMTREC's teleconferencing bridge, the shipper's emergency response coordinator can be linked with the participating team's contact to coordinate the response.

At the scene, the team(s) will provide a chemical industry presence and may advise the incident commander on safe techniques for sealing containers, fire and explosion prevention, spill mitigation, and coordination of cleanup activities. In most cases, the teams are equipped to provide "hands-on" assistance such as capping/patching containers and in transferring products to secure containers or provide any other assistance as requested by the on-scene incident commander.

CHEMNET utilizes the considerable expertise and emergency response capabilities of both the chemical industry and private contractors. This network provides coverage to the entire country and is not unique to certain products or geographic areas, and it complements existing mutual-aid agreements sponsored by makers of certain groups of chemicals, such as CHLOREP and PERT.

Membership CHEMNET membership is open only to CMA members and nonmembers who are chemical producers and who are CHEMTREC registrants at no cost. Participation in CHEMNET is a valuable asset for those chemical producers who do not have emergency response capabilities because of size or other reason. The network enables these producers to use an extensive and efficient emergency response program at minimal cost.

Chemical Referral Center
The Chemical Referral Center (CRC), at (800) 262-8200, is designed to help public transportation workers and users of chemicals obtain health and safety information about chemicals and chemical products. The center refers nonemergency inquiries to participating chemical companies.

The center operates at CMA headquarters in Washington, D.C., from 9:00 a.m. to 6:00 p.m. (EST), Monday through Friday. When an inquiry comes into the center about a chemical, the operator first determines that it is not an emergency (a spill, leak, fire, explosion, or accident). All emergencies are immediately rerouted to CHEMTREC. If the inquiry is not an emergency, the operator finds out the name of the company that manufactures the product in question. Working from a computerized index that contains information on the manufacturers of chemicals, the operator supplies the caller with the address and telephone number of the company to call, along with a contact.

CMA Lending Library
The CMA Lending Library is an exhaustive resource that details many reviewed audio-visual training programs for hazardous materials emergency response personnel. The list is constantly being reviewed and updated. A current copy of the list may be obtained by contacting the CMA at (202) 887-1216 or by writing to:

NCRIC
Chemical Manufacturers Association
2501 M Street, NW
Washington, DC 20037

Emergency Response Training
The CMA offers access to regional workshops for emergency response personnel. These courses, normally 3 days in duration, are peer-reviewed prior to their presentation and are taught by current industry professionals. Courses are given several times per year at various locations across the United States.

CAER/TRANSCAER
CAER/TRANSCAER stand for Community Awareness and Emergency Response, and Transportation Community Awareness and Emergency Response. CAER/TRANSCAER are voluntary programs developed by the CMA to inform local residents of industry operations in and transportation routes through their community. Additionally, the programs serve to integrate industry and community response mechanisms, and to test and modify these plans through regular exercises.

An interesting development that parallels and somewhat follows the CMA's initiative is the establishment of local emergency planning committees (LEPCs) under the Emergency Planning and Community Right to Know Act of 1986 (SARA Title III). Members of CAER/TRANSCAER are fully committed to integrating their voluntary functions with LEPCs.

2

Crisis News Management for Hazardous Materials Incident Responders and Incident Managers

Gordon Norman

To refer to the events surrounding a hazardous materials incident as a "crisis" is really an understatement. A "crisis" is defined as "a decisive moment," "a turning point." For most companies, what actually happens immediately following a hazardous materials incident would be more correctly defined as "chaos," "a state of unorganized confusion."

Cleaning up the incident site is often the least of the company's problems. In fact, most companies either have, or have access to, well-trained and effective response teams. What creates the corporate confusion is management's inability to deal effectively with the public and media concerns that surround the incident. Corporate managers are not generally well skilled in media news management and public relations; consequently, when a crisis occurs, they attempt to deal with it in a disorganized, ad-hoc manner.

The public response following an incident is charged with misunderstandings, mythologies, fears, and anxieties. The media share these public anxieties. The company responsible for the incident is not playing on a level field. All the parties involved approach the issues with preconceptions that impede communication. Understanding these preconceptions is an essential prerequisite to effective news management.

1. THE NEWS MEDIA AS ENTERTAINERS

The news media are more than public information agencies: They are active participants in the global entertainment industry. Media success is measured by the size of their audience. Most of the information that fills the spaces between the advertising in newspapers and most of the time between the commercials on radio and television is carefully structured to appeal to a mass audience—an audience for whom the passions and fears associated with sex, violence, war, crime, and large-scale disasters are major areas of interest. News stories that focus on these topics attract large audiences and help sell the news media's product. Make no mistake about it: The news media are in business; timely, entertaining information is the product they are marketing.

Most news editors try to maintain a balance between pandering to the passions and fears of the masses and providing accurate information about the events of the day. They do not claim to be objective or unbiased; they do however, try to be fair and to present the public with a variety of viewpoints on the issues. The media are constantly subject to public scrutiny and therefore they strive to avoid unfair and misleading statements or errors that would affect their credibility and the sale of their product.

The most frequently heard criticism of the press is that they are interested only in sordid and sensational new items. This is not true. All media, whether publicly or privately owned, are interested in "any" news item that they feel will interest their audience. The media interest in hazardous materials incidents is generated by their audience. The sensationalism surrounding such incidents usually results from some misdeed or mishap that is caused by those who are manufacturing, handling, or regulating the hazardous materials. The media cannot be faulted for covering such incidents.

2. THE NEWS MEDIA AS ENVIRONMENTAL OPINION MAKERS

The media not only cater to the public interests, they help to create those interests. The popular interest in environmental issues is a case in point. A great deal of the public obsession with the environment is attributable directly to the widespread media coverage of such issues. Media coverage has helped create a public so sensitized to environmental issues that people now view every hazardous materials incident as one with negative global implications. To this public there are no "insignificant" hazardous materials incidents. In their eyes, companies who are involved in hazardous materials incidents are assumed to be guilty of environmental crimes until proven innocent.

Safe, wholesome environmental stewardship is rapidly becoming the religion of our time. The earth, the sea, and the air form the holy trinity of a new environmental fundamentalism. Those who pollute this trinity are guilty of the gravest sacrilege. To the environmental extremists of the new religion, there are no acceptable compromises. Fortunately, the extremists are in the minority; the majority of the public and the media are still open to compromise.

A company that has just dumped a tankcar load of vinyl chloride on a city street would be foolhardy to condemn the public and media concerns by suggesting that they are overreacting or that they are being mislead by environmental extremists even if that were true. The validity of the public's opinion is not the issue. We may indeed be a nation of environmental hypochondriacs, oversensitized to the environment's needs and overcritical of those who manufacture and handle hazardous materials. Company officials and incident responders must deal with the public perception as it is, not as they might wish it to be. The public perception, right or wrong, is the reality that must be dealt with.

3. THE GROWING PUBLIC DISCONTENT WITH MODERN BUSINESS ETHICS

Even before a hazardous materials incident happens, the company that is responsible has several strikes against it, not the least of which is that the public are predisposed to distrust the business community in general.

There is widespread public discontent with corporate behavior. Every day, news reports provide evidence of fraudulent business practices, bureaucratic coverups, corporate environmental misdeeds, stock and bond manipulations, pricing scams, and a host of other unseemly business activities. It is no wonder that the public perception is that the majority of businesspeople are dishonest and greedy. The critics of modern business behavior have more credibility than the business leaders themselves. No business is immune to the criticism. The ethical foundation of business itself is under attack. Companies that are involved in hazardous materials incidents can expect to be subjected to the merciless and critical scrutiny of a disbelieving public and media.

Furthermore, the hierarchical structure of many corporations predisposes management to take a defensive posture when a crisis occurs. The *modus operandi* of some company managers is to cover up any responsibility for the incident. Credible deniability is the underlying objective of those managers who, like Oliver North, attempt to rationalize the misdeeds of their superiors—a bureaucratic knee-jerk reaction that only exacerbates the public distrust of the companies and organizations involved.

Neither superficial public relations ballyhoo, corporate advertising hype, legal mumbo jumbo, or technical double-talk will ease public anxiety when a hazardous materials incident occurs.

Restoration of the public trust can be achieved only when management successfully addresses the ethical and philosophical issues that underlie the public predisposition to distrust the company. Corporate policy makers cannot evade these issues by hiding behind the image-oriented defensive tactics of their advertising and marketing departments. Nor can they expect their hazardous materials response personnel to handle effectively all media and public enquiries that follow from an incident. They must assure the public that the economic dynamic of their business is not incompatible with sound environmental stewardship, or, failing that, they must be able to convince the public that their business activity, while endangering the environment, is a justifiable compromise.

Responding to a critical attack on the company's values is not the job of incident responders; it is a task for senior management. Management strategies and tactics designed to deal with these issues are the subject of other books. The issue has been raised here only to illustrate that when an incident occurs, the company is faced with a cynical public that has already reached a guilty verdict—a public that searches the media news reports for evidence that the guilty verdict is justified.

The tactics that follow are designed for use primarily by incident responders, incident command teams, and their supervisors, who should not be expected to publicly defend or philosophize about the state of the company's ethical values. Incident responders who are being interviewed by the press should focus their answers on incident-related data only and leave the ethical questions to corporate policy makers.

4. WHAT TO EXPECT WHEN AN INCIDENT OCCURS

4A. The News Media Will Hear About the Incident

No company should attempt to keep a hazardous materials incident secret. Those who try it may find themselves in a situation where company personnel are denying that any incident has taken place to one group of news reporters while at the same time the local fire chief is giving details of the incident to another group of reporters.

Most media newsrooms monitor police and fire department radio communications. Programmable radio scanners also permit news reporters to "listen in" on response team and interagency communications. Companies and response agencies should never assume that their radio frequencies are secure from the ears of the media.

4B. The Media Will Expect Access to Information

It is important to make a distinction between "access to information about the incident" and "physical access to the incident site." News reporters will want both. Incident responders have a responsibility to restrict access to any area where there is a danger that the public may be harmed. Reporters, photographers, and camera crews should not be permitted to wander about in areas that are dangerous. Incident responders have both a right and an obligation to prevent access to hazardous areas.

Access to information about the incident is a different story. The public has a right to information. Hazardous materials incidents that endanger the health of the public or threaten the environment are "public incidents." Those who manufacture or handle hazardous materials have an ethical obligation, and in some jurisdictions a legal obligation, to provide the threatened community with such information as is required to make informed decisions concerning health and safety.

Incident response agencies and companies involved with the cleanup risk public condemnation when they instruct employees to give only their name, rank, and serial number to journalists who are seeking incident-related information. Journalists who encounter a wall of silence may assume that the company has something to hide. "No comment" policies only exacerbate what is already a public relations disaster and could lead the media to conduct an in-depth search for information that they imagine you are hiding.

Incident commanders should assign at least one person to assist the media at the incident site. Companies that manufacture or handle hazardous materials should have corporate background information available in advance of an incident. Corporate media information kits are an effective means of providing reporters with accurate information about the company. These media kits should be available to the press at the incident site.

4C. The News Media Are Not "Out to Get You"

Having conducted dozens of seminars on crisis news management, I am amazed at the number of company executives who actually believe "the press" is out to get them. This simply is not true. News reporters are not out to get anyone or any company. They are out to get the story. Most reporters are busy professionals and have neither the time nor the inclination to devote to personal vendettas against companies that handle hazardous materials. In many cases, when reporters are dispatched to the scene of a hazardous materials incident, they already have a full day's work planned; their prime objective is to get the facts about the incident as quickly as possible, file the story, and get on with their regularly scheduled assignments.

News reporting is a time-sensitive activity. Radio and television reports may be broadcast live from the scene. These journalists are often working toward deadlines that are only minutes away. They want accurate information and they want it fast.

The constant pressure of impending news deadlines can cause reporters to behave in an aggressive, impatient, and arrogant manner. Don't get involved in an argument with these reporters. When dealing with them, it is best to say what you have to say as succinctly as possible. Stick to the known details. Never say "no comment" without a clear explanation of why you are saying it. If you do not know the answer to a question, direct the reporter to someone who does. Remember that all questions about corporate ethics and the company's policies should be dealt with by company executives.

4D. The Media's Questions Will Fit the Media's News Formula

Journalistic behavior is generally predictable. News reporters approach most assignments with the same standard set of questions:*what, where, when, who, why*. For example: What happened? Where did it happen? When did it happen? Who was hurt? Why did it happen?

There are endless variations to this standard news formula. A reporter might ask: What is the chemical? Where does it come from? When did you acquire it? Who is in charge of it? Why was it stored here?

Of course, the first question that news reporters will ask is "Who is in charge here?". If you happen to be the person in charge, you should be prepared to answer any of the above questions.

Anyone working at the scene of an incident may be questioned by the press. No one is immune. In covering large incidents, media personnel may question everyone involved, from the gatekeeper to the company president. To a news reporter, anyone working at the scene of a hazardous materials incident may be considered a "spokesperson."

4E. Never Underestimate the Power of the Press

The "on-the-scene interview" is only a small part of the journalistic activity that focuses on the incident and those involved. The information gathered at the scene will be assessed and processed in any, or all, of the following ways:

- The information will be compared with information from other sources. For instance, if a company spokesperson has told a reporter that the spilled chlorine will not pose a long-term hazard to the environment, the reporter will certainly have the statement confirmed or denied by a recognized authority on chlorine.
- The press will use their company files to acquire background information about the companies involved in the incident. Most journalists have access to prior news stories. Errors that have appeared in past news stories are likely to be repeated.
- News reporters will share information with other media personnel. It is unwise to show favoritism in releasing incident-related information. You can't get away with giving information only to reporters who are sympathetic to your cause while withholding the same data from reporters who are hostile. Companies who try this discover that they have only created more hostility.
- When media executives determine that the incident justifies the costs, they have access to a vast array of technical resources. It is not uncommon for the media to employ helicopters to film an incident site. They also have access to a full complement of high-tech listening devices, long-range optical lenses, and a trained research staff.

Intensive investigative journalism is rarely practiced: It is simply too costly for most media organizations. Consequently, such investigative activity is reserved for major stories such as the Three-Mile Island incident or the *Valdez* oil spill. News editors will not initiate a costly investigation unless they are certain that the story is big enough or spectacular enough to justify the time and money.

More investigative journalism is seen in TV soap operas than in the real world of journalism. Unless a company is rife with corruption and engaging in criminal activity, it ought not to fear the investigative powers of the press.

4F. Using the News Media as an Ally

Incident commanders should not hesitate to ask media reporters for help. The news media will often withhold or delay incident-related information that is harmful to the public if they are asked to do so. For example, news editors regularly withhold the names of victims until their families can be notified. Film clips of cars involved in tragic accidents are frequently edited to hide the licence plate numbers in an effort to conceal the identities of victims. Sometimes the exact location of an incident will not be released until police and incident responders have indicated that the area is safe.

For instance, a news report might indicate that "a truck carrying fireworks is burning in the north end of the city." By not giving the exact location, the reporter avoids turning the incident site into a destination for sightseers. The exact location can be given in later news reports, after the event is over.

When asking the news media to withhold or delay information, keep in mind that they will do so only if they are convinced that the public interest is being served.

The news media can also help by publicizing the addresses and phone numbers of incident-related services—for example, temporary phone numbers for employees, families of victims, etc.

The media attention focused on a company during and after an incident can be seen as an opportunity to promote the company's image. The media will be looking for "human interest" stories related to the incident. Response agencies and companies involved can use this opportunity to provide the media with human interest stories that project a desirable corporate image. There is no guarantee that such stories will actually be used, but the media are not immune to a little gentle manipulation by skilled news managers. Local newspapers have a constant need to fill the spaces between the advertisements, and they are often willing to use company-prepared news materials.

4G. Hazmat Jargon and Technical Information

It is likely that the news reporter has arrived at the scene of the incident without any technical background information. News reporters are generalists; they are unfamiliar with the technical terminology and jargon used by incident responders. Incident responders should take care to explain the technical details about what they are doing so that news reporters have a clear understanding of the events that are taking place. Incident responders and company representatives have the information needed to ensure that news reports are accurate. They should see to it that news reporters are briefed accurately.

All technical descriptions must be understandable to the general public. Even such common technical terms as "oxidizers" and "catalysts" should be clearly defined. When describing industrial chemicals, indicate what the chemical is used for. If possible, give an example of a common household use.

Never downplay the dangerous nature of a chemical product. An inaccurate or misleading description of a hazardous material will seriously affect the company's credibility and could have legal implications that cause the company irreparable damage.

5. THE "ON CAMERA" INTERVIEW

5A. Facing the Cameras: How You Look and What You Say

When conducting media seminars for incident responders, the most frequently asked question is whether one should look directly at the camera when being interviewed. There's a simple solution. If peering directly into the camera lens makes you feel uncomfortable, don't do it. The alternative is to face the reporter and let the camera operator figure out how to adjust for the camera angle. While you are being filmed, you should stand in one spot without moving about.

The camera operator will want to film the interview from a position that shows a view of the incident scene in the background. Don't let the media choose the background scene for any on-camera interviews. Remember that you are in charge, not the camera operator. The key things to avoid are scenes in which the background events could detract from what you are saying. Watch out for inappropriate billboards on nearby buildings and protest signs held by picketers at the scene. Conducting the interview in front of or alongside a fire engine or response vehicle will usually provide the media with a suitable background view.

5B. How You Look

When the news item appears on TV, your appearance may be as important as what you are saying. Viewers can be easily distracted or offended by the appearance of people they see on TV. It is important that your dress and demeanor be appropriate. Don't wear sunglasses, and don't smoke or chew gum while being filmed.

Don't peer over your glasses at the camera, and remove flashy jewelry before the interview. Don't allow yourself to be interviewed while sitting behind a desk or while standing in the company's board room. The public are not impressed with displays of wealth or symbols of corporate power. Keep in mind that the public are preconditioned to distrust the company. Dress accordingly.

Incident responders will probably be in uniform when they are interviewed at the scene of an incident. Safety helmets can cast shadows across the face. Those who wear safety helmets should either remove them or push them back so that the face is not obscured when being filmed.

Response personnel should be very cautious about wearing T-shirts that carry corporate logos or messages that do not apply to their employer or the work they are doing. T-shirted employees who are walking billboards for beer companies, vacation resorts, or gun clubs are not dressed appropriately. To illustrate this, imagine an incident where a school bus full of children has just collided with a truck carrying chlorine. The chlorine tanks have ruptured, and several children have been killed. An incident responder wearing a T-shirt with a military cartoon and the caption "Kill 'em all—Let God sort 'em out" steps before the camera to answer a reporter's questions. Scenes like this can cause irreparable damage to the company's image. Even less offensive messages on T-shirts are inappropriate at the scene of an incident and should not be worn while on duty. Baseball-style caps with inappropriate logos or slogans printed on them should also be avoided.

5C. What You Say

a. If You Are a Company Spokesperson

Company spokespersons will undoubtedly be met with questions about company policies and company ethics. Considerable diplomacy is called for. The more senior your position in the company, the more likely you are to be asked questions about these issues. When answering media questions, be careful to address your answers to all of the company's stakeholders. Don't overlook the shareholders, the employees, the customers, or the general public. For instance, you might be asked, "How long will your plant be shut down?" A reply stating that you "will stay closed for as long as it takes to ensure that the workplace is safe for the employees" may sound just fine to some, but it leaves a lot of questions open to speculation. Employees will want to know if they will be paid during the shutdown. Clients will want to know if their orders will be filled. Local businesses that depend on the company will also not be comforted by the prospect of a prolonged shutdown. Company creditors will wonder if the shutdown will affect your ability to cover current liabilities, and shareholders will consider dumping their stock. The statement fails to address all of the company's stakeholders.

As a company spokesperson, whatever you say will be considered by the media as the "official company position." Since incident-related news media questions are predictable, it is not difficult for company spokespersons to prepare their answers in advance of the interview.

The media will expect the company to hold a press conference as soon as possible after the incident. The press conference venue makes it easier for the company's senior executives to work as a team and avoids having a single executive put on the media "hot seat." Companies involved in large incidents make a serious mistake when they attempt to manage the news media through their advertising and public relations departments. If the incident is a very large one affecting a wide area, it is advisable to employ a professional news manager who has experience in dealing with hazardous materials incidents.

b. If You Are an Incident Responder

Having just read the above advice for company spokespersons, you can see that they are the ones most likely to get the "hot seat" treatment from the news media. Incident responders who are asked questions about the company's ethics, company policies, or issues that relate to the company's commitment to the community should direct the reporters to the company's executives. This is one time when it is prudent to pass the buck.

The irony is that responders are usually better trained to handle media interviews than company executives are. Most incident responders have taken "hazmat" response training courses that include training in press relations and news interviews.

Media journalists and the general public tend to view incident responders as part of the solution and executives of the companies involved as part of the problem. This distinction is an important one. Incident responders enjoy more public credibility than do company executives. Responders are expected to be at the incident site to clean up the mess, not cover up the mistakes and misdeeds that caused it.

Incident responders should restrict their comments to incident-related activities. They should not attempt to answer any speculative questions about "how" the incident occurred or "why" it occurred. Focus on details about the cleanup and avoid details about the cause.

Incident commanders and responders should provide the press with a clear description of the problem and indicate what is being done to clean it up. Incident response teams are engaged in a dangerous business, and their lives are often at risk. They have no reason to fear the media. Every reporter knows that if the incident calls for heroic deeds, it will be an incident responder who rises to the challenge.

6. A NOTE ABOUT MEDIA KITS

Media kits are important public and media relations tools for companies that handle hazardous materials. Prepared in advance, effective media kits contain a wealth of information for use by reporters.

The kits should contain a short history of the company and an overview of the company's policies that relate to the environment and the community at large. They should also contain a list of key personnel, phone numbers of company contacts, selected maps of the company's property, and any other company details that they feel would be of interest to the media. When an incident occurs, detailed information about the chemicals and products involved should be added to the media kit.

The media kits should not to be confused with the company's contingency plan. Kits should be designed specifically for the media and should be available for distribution at the incident site. Great care should be taken in producing them. Every statement in them will be subject to thorough scrutiny. Media kits are not advertising brochures or corporate image promotions. They should not be written from a marketing perspective. They are documentary statements about the company, its management, its employees, its products, and its relationship to the community.

New reporters who are provided with effective media kits are much less likely to pester incident responders with questions about the company. The kits are a company's best defense against media inaccuracies and unfounded speculation about the company's environmental policies.

3

Personal Protection Equipment

This chapter discusses three aspects of personal protection: respiratory protection, protective clothing, and decontamination procedures.

1. RESPIRATORY PROTECTION

A wide variety of respiratory protective equipment is available to incident responders; different manufacturers have developed features and techniques exclusive to their product. The aim of this section is not to critique the values or shortcomings of the various commercial types, but rather to show the generic types available and compare their uses in various incidents.

Two broad types of respiratory protective equipment are generally available: atmosphere supplying and air purifying. Within these two categories several different types of equipment are available. Figure 3.1 shows, in diagrammatic form, the types of units available.

1A. Self-contained Breathing Apparatus

Figure 3.2 illustrates a typical design for a self-contained breathing apparatus (SCBA). A supply of air is held in the pressurized air cylinder, the exact fill pressure depending on the intended duration of use and cylinder design. Fill pressures may range from 15 158 kPa (2200 psi) to 31,005 kPa (4500 psi). This air then passes through a pillar valve on the cylinder to a first-stage reducer, which lowers the pressure to approximately 551 kPa (80 psi). From the reducer valve the air is fed directly to an admission valve, which feeds to a face piece. Depending on the type, this admission valve is located either on the face piece or on a chest-mounted control

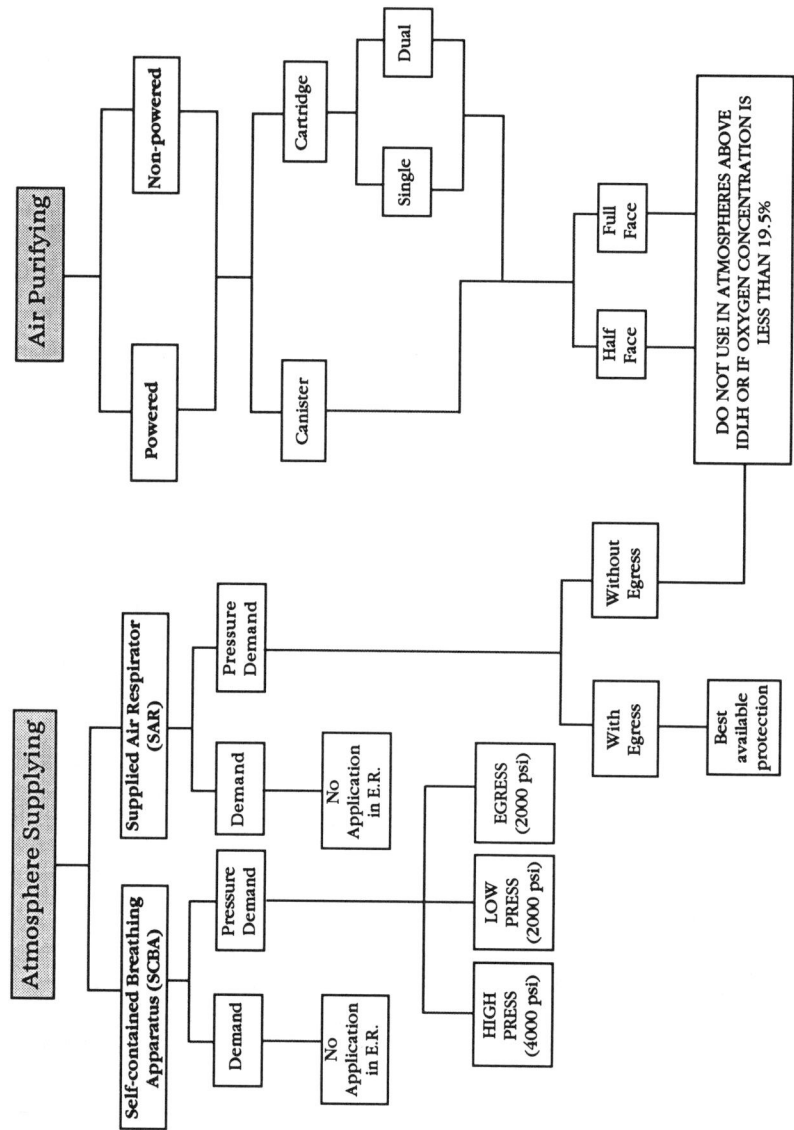

FIGURE 3.1. Respiratory Protection Equipment for Spill Response

1. Respiratory Protection   39

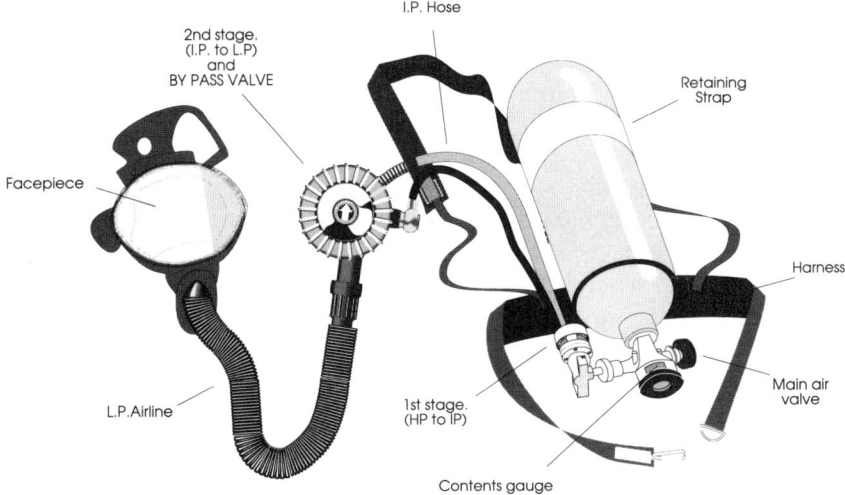

FIGURE 3.2. Two-Stage SCBA   (Courtesy of SURVIVAIR)

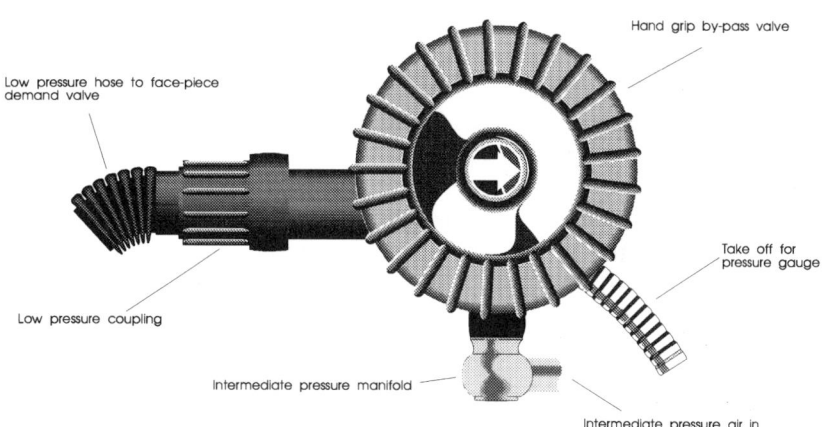

FIGURE 3.3. Admission Valve   (Courtesy of SURVIVAIR)

unit. Figure 3.3 shows a typical admission valve design as found in units where the reducer valve is located on the wearer's chest.

In the past, units were designed to be either pressure demand or demand operated. The essential difference between the two is that on the pressure-demand units a small positive pressure, approximately 6.89 kPa (1 psi), is always maintained by means of a spring in the diaphragm and an exhalation valve that is set above atmospheric pressure.

Demand-type SCBAs rely on the wearer creating a pressure below atmospheric to cause airflow, this being achieved by inhalation. This action, however, may cause an ingress of the external atmosphere, thus defeating the purpose of a supplied-atmosphere unit. Those units that are currently in use are being retrofitted, and all new SCBAs must be of the pressure-demand type.

All SCBAs are fitted with a bypass valve designed to circumvent the operation of the reducer valve in the event of an emergency. It cannot be overemphasized that this is an *emergency valve*. Whenever its use is dictated (e.g., because of suit failure or air starvation), the next step, after opening the valve, is to leave the hazardous area. In other words, this valve is not designed to increase face pressure for wearer comfort or for use as a defogging valve.

Other design features of SCBAs aimed mainly at safety include pressure gauges at the air cylinder pillar stem and at a point observable by the wearer, and a low-level alarm which activates when the bottle contents are reduced to 20–25 percent of full capacity.

Before contemplating the use of a SCBA, or any other respiratory protective apparatus, the following criteria should be assessed:

- Have intended wearers been fully trained in the use of SCBAs, and have they passed a qualitative or quantitative fit test?
- Is the unit resistant to all vapors or liquids to which it will be exposed? If not, ensure that a totally encapsulating suit is used.
- What is the distance from the support zone to the exclusion zone, and how much time will it take to reach the incident/leak area? This is a critical factor in deciding the distance between a command post or staging point and the exclusion zone. While it is important to stage operations at a safe distance from potential exposure, it is equally important that entry teams have to walk a minimal distance to approach an incident, in order to ensure maximum duration of available air supply.
- Can wearers be observed from the support zone at all times? This may be a difficult objective to achieve, especially when incident resolution requires working in areas that are out of sight of the command post. If wearers cannot be observed constantly, it may be necessary either to relocate the command post or to stage an observation team at an intermediate point.
- How much time will be necessary for decontamination prior to permitting the wearer to breathe fresh air? Decontamination processes may take anywhere from 30 seconds to several minutes. An allowance for the decontamination process should be included in the anticipated duration of use.

- What is the wearer's anticipated bottle consumption time? The time taken to consume the contents of an air bottle varies significantly with the individual. Prior to an incident occurring, all team members should wear an SCBA in the prescribed level of protective clothing and undertake a level of exertion that approximates response activities. This pre-incident testing provides an approximate anticipated bottle duration for each individual.

The above factors can be combined to give an assessment of the amount of time available in the exclusion zone by using the formula

$$\text{Time (in zone)} = \text{evaluated bottle duration} - (\text{suit donning time} + \text{approach time} + \text{return time} + \text{decontamination time} + 25\% \text{ of evaluated bottle duration})$$

This formula demonstrates that actual work time in the exclusion zone is considerably less than manufacturers' nominal duration.

Other factors to be assessed include the following.

- Would other methods of respiratory protection be more appropriate? Alternatives to consider include a supplied-air respirator with egress (SAR) or a closed-circuit breathing apparatus. This is especially so if the required duration at the work area cannot be achieved.
- Have clear communications procedures been established between the entry party and the support group and within each group? Without such procedures, safe, effective control of an entry cannot be achieved.
- Are emergency procedures clearly established? The buddy system must be used at all times during incident management. For every person who enters the exclusion zone, there should be a backup person, similarly equipped, available in the support zone. In other words, for safe approach, a minimum of four equivalently equipped personnel is required. Should an emergency occur involving the entry team, a backup is immediately available.

Other issues pertaining to incident approach are addressed in Chapter 4.

While the exact operation of an SCBA varies from type to type, the following procedures should be observed prior to and during the operation of a unit.

- While the unit is still housed, check that the pillar, main, and bypass valves are all closed. Check that the face piece and its associated hose are clean and free of debris. Read the pillar valve content gauge to ensure that there is sufficient air for the intended task. Check that all harness

straps are fully slackened, buckles are opened, and the harness is in good condition.
- Formal fit testing will have been carried out sometime prior to wearing a unit. Whenever a unit is to be worn, the following " quick fit test" should be carried out. Place the face piece in position and inhale through the nose. It should be possible to hold the face piece in position by this action. If a vacuum is not maintained, then either the face piece or its piping are leaking, or the face piece is not suitable for the intended wearer. Failure of this test indicates selection of another unit.
- Open the cylinder pillar valve. This will cause a flow of air from the face piece (due to the pressure-demand design) and pressurization of all components. Check other parts of the assembly and couplings to ensure that they are not leaking. Compare the pillar and remote pressure gauges and ensure that they indicate the same value.
- Briefly open the bypass valve to check its operation. Then ensure that it is securely closed.
- Close the pillar valve and slowly "breathe" the unit down to zero pressure. Ensure that the low-level alarm actuates at approximately 25 percent of working pressure.
- Don the unit and secure all straps to ensure comfort. Tilt the head back as far as possible and ensure that contact is not made with the air bottle.
- Reopen the pillar valve, and ensure that it is fully open and locked in position.
- Don the face piece. The unit is now ready for use.

During use, the wearer should constantly monitor the unit's operation and the level of remaining air. In case of air flow restriction, alarm activation, discomfort, or other untoward conditions, the wearer should immediately leave the exclusion zone.

After use, the unit should be thoroughly cleaned and inspected, and the face piece sanitized and placed inside a plastic protective bag. Ensure that the air bottle is full and that all components are in working order, the pillar valve is closed, and all hoses are depressurized; in this way the next wearer can use the unit with confidence.

Table 3.1 describes the features, advantages, and disadvantages of SCBAs.

1B. Supplied-Air Respirators

Supplied-air respirators (SARs) often present a satisfactory alternative to SCBAs. Their main advantage over SCBAs is their light weight and extended duration capabilities.

1. Respiratory Protection   43

TABLE 3.1   SCBA Features

Advantages	Disadvantages
Mobility	Weight
Safe for use in atmospheres above IDLH	Shorter duration
Safe for use in oxygen deficient atmospheres	Re-charge requirements
Positive pressure	Initial cost
No external management	Awkward in restricted spaces
No "umbilical"	

Figure 3.4 illustrates a typical supplied-air respirator design. The arrangement consists of a bank of compressed air cylinders that feed a distribution bar. Air is fed to the bar by way of a pressure reducer that ensures a constant delivery of approximately 551 kPa (80 psi) at the bar. This air is then fed through a durable, lightweight, chemical-resistant hose to a face piece. The length of this hose is generally restricted by regulatory agencies to 300 ft. The face piece is fitted with a diaphragm-type admission valve.

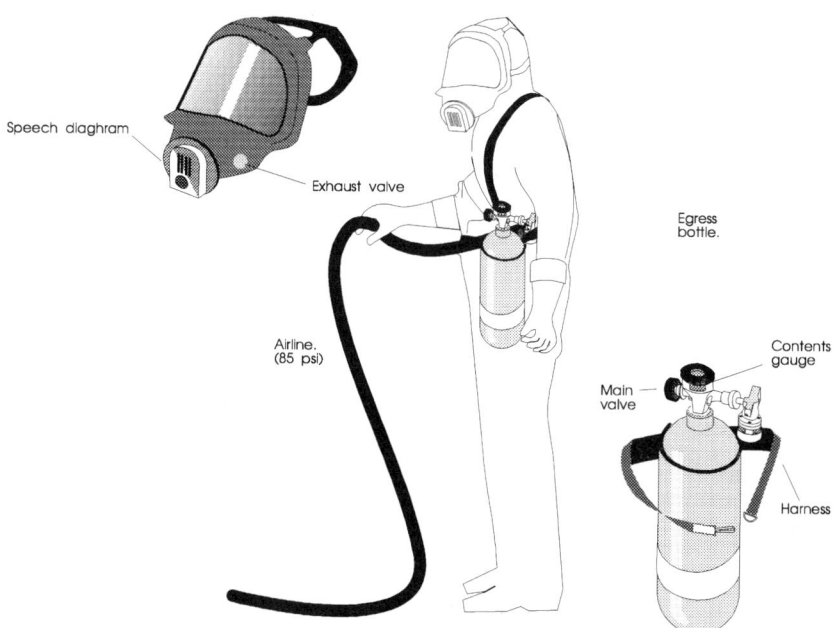

FIGURE 3.4.   Supplied-Air Respirator with Egress

Like SCBAs, these units are designed to work on pressure demand. Obviously, should the air supply fail due to severing of the hose, entrapment, kinkage, or similar disruption, the wearer will be deprived of air. In light of this, these units should be operated only in conjunction with an emergency air supply. In fact, in atmospheres where the IDLH (immediately dangerous to life and health) value of a product is exceeded, or where the oxygen level is less than 19.5 percent, such an emergency air supply is required by regulations. This supply normally takes the form of a small, hip-mounted bottle of compressed air. In the event of air line or supply failure, the wearer opens the bottle pillar valve, which automatically isolates the air line feed, disconnects from the airline, and leaves the exclusion zone. This equipment is known as an egress unit.

Before using a supplied air respirator the following criteria should be assessed.

- Are the face piece unit, hose, and egress unit all resistant to degradation by vapors or liquids to which they will be exposed? If not, ensure that a totally encapsulated suit is used and that a chemical-protective overhose is fitted to the supply air hose.
- Can the incident area be reached with hose to spare for maneuvering around the site? Remember, the maximum permissible hose length is 300 ft. Once an entry party is in the exclusion zone, extra hose length will probably be needed in order to safely reach valves, containment devices, and other such apparatus at the site. Should it prove necessary to work above ground level, sufficient slack hose will be necessary to ensure a vertical drop from the wearer to the ground. Any horizontal tension will tend to cause loss of stability of the wearer and heighten the risk of slipping or falling.
- Can wearers be observed by the command team at all times? The same arguments presented before with respect to use of SCBA apply here.
- Are liquid pools covered to ensure that the hose does not lie in contaminants? If not, some manufacturers can supply chemical-resistant overcovers for the hose. Ideally, the entry team should carefully lay out their air hose in areas that will avoid contact with liquids. These precautions reduce the risk of hose contamination or degradation from contact with contaminants.
- Are communications and emergency procedures clearly established (similarly to those discussed with regard to SCBA).

After use, all equipment should be decontaminated, cleaned, and sanitized as discussed previously. Particular attention should be paid to the cleaning and condition of the air line. Table 3.2 details the advantages and disadvantages of supplied-air respirators.

TABLE 3.2 Supplied Air Respirator Features

Advantages	Disadvantages
Duration at incident site	Hose length (300' max)
Light weight	Needs egress for IDLH or O_2 deficient atmospheres
Moderate cost	Needs trained operator for cascade
Pressure demand	Cascade bottles are heavy

1C. Air-Purifying Respirators

Air-purifying respirators (APRs) and powered air-purifying respirators (PAPRs) come in a variety of configurations, i.e., full face piece, half face piece, nose cap, single cartridge, dual cartridge, canister, etc. Some of these types are illustrated in Figure 3.5.

The essential difference between APRs and PAPRs is that a PAPR uses some form of fan/blower-type unit to give a positive air flow at the wearer's face. This gives some degree of positive pressure along with, for some, an increased degree of comfort. In no way does this positive pressure allow the unit to be operated in an IDLH atmosphere.

Both units work on the principle of filtering particulate out of untreated air or absorbing contaminants into an appropriate medium. The filter/absorption material is contained in either a cartridge or canister. Generally, canisters offer a longer duration of exposure. As a general guide, these units are designed for wearer comfort more than for total protection.

a. Limiting Factors

All cartridges are rated for protection against specific substances. Depending on the type, this may be one or several such substances; the manufacturers' data sheets and the cartridge labeling will clearly indicate the substances against which the cartridge is suitable for protection. The limiting factors for use of air-purifying respirators are:

- The atmosphere must be constantly monitored for oxygen concentration and contaminant levels. All airborne contaminants must be clearly identified.
- A minimum oxygen concentration of 19.5 percent must be maintained at all times.
- Unless specific regulatory approvals allow otherwise, APRs may be used only for protection against gases with an odor threshold below the permissible exposure limit (PEL).
- The manufacturer's maximum rating must not be exceeded.
- The concentration of airborne contaminants must not exceed the IDLH value.

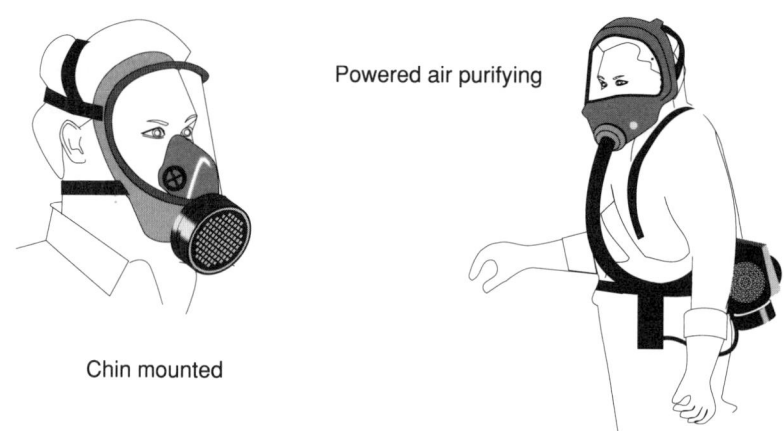

FIGURE 3.5. Air-Purifying Respirators

- The assigned protection factor × PEL must not be exceeded.

The safest approach is to use APRs and PAPRs for comfort only, in a fully characterized atmosphere. Should the unit fail, the wearer must be able to walk out of the contaminated area while experiencing only mild discomfort. In spill response situations it is often difficult to establish exact atmospheric characteristics in all areas. Bearing this in mind, great

1. Respiratory Protection 47

TABLE 3.3   Advantages and Disadvantages of Air Purifying Respirators

Advantages	Disadvantages
Small, lightweight	Not suitable for atmospheres above IDLH or deficient in oxygen
Low maintenance	
Low initial cost	Cartridges can be expensive
Little operator training required	Must constantly monitor atmosphere
Least movement restriction	Seal can break down during inhalation

consideration must be given prior to selecting an air-purifying unit in preference to an atmosphere-supplying unit.

b. Duration of Use
The anticipated duration of use and the shelf life of a cartridge are indicated on the manufacturer's label and data sheets. When a cartridge nears the end of its life, either air flow will become restricted (in the case of a particulate filter-type cartridge) or odors of the substance being protected against will be detectable (in the case of an absorbtion-type cartridge and an odiforous material).

c. Reuse of Cartridges
Prior to use, a cartridge should be kept in a sealed, tagged, plastic bag. If it is not, do not use the cartridge. After use, cartridges should either be disposed of or sent for decontamination, flow testing, and rebagging. In some circumstances, cartridge decontamination may not be a practical procedure, for example, when it has been used for a PCB cleanup operation.

Table 3.3 illustrates the advantages and disadvantages of air-purifying respirators.

1D. Common Considerations

There is one final consideration that is common to all types of respiratory protective equipment. This is:

a. Protection Factors
All respiratory protective equipment is assigned a *protection factor* that relates to the ability of the equipment to protect the wearer. There are several types of protection factor (assigned, simulated workplace, and workplace); the most often quoted is the assigned protection factor (APF). This gives a measure of the minimum anticipated level of protection.

TABLE 3.4 Summary of Assigned Protection Factors

Respirator Type	Assigned Protection Factor
Half Mask APR	10
Full Face APR	50
Half Mask PAPR	1000
Full Face PAPR	2000
SCBA/PAPR (pressure demand)	10000

Protection factors are one criterion used in deciding whether an apparatus is suitable for use in a contaminated atmosphere; one ceiling factor is the PEL × APF. Assigned protection factors for the different types of respiratory protective equipment are shown in Table 3.4.

As an example, when dealing with toluene (PEL = 100 ppm), one of the limiting concentrations for a full-face-piece APR would be 5000 ppm (PEL × APF). Note that in this example the IDLH for toluene (2000 ppm) is exceeded and now becomes the limiting factor. If a half-face-piece APR is used, then the limiting value is 1000 ppm (PEL × APF).

2. PROTECTIVE CLOTHING

Most materials encountered at hazardous materials incidents will present the risk of some degree of harm in the event of an exposure, whether from liquid contact or exposure to harmful vapors. The principal objective of this section is to illustrate and compare the various levels and types of protection available and to discuss their limitations in use.

2A. Levels of Protection

There are four generally accepted levels of protection afforded by protective clothing, levels A, B, C, and D, where Level A offers the highest level of protection. This nomenclature, which is promulgated in the National Fire Protection Association's NFPA 471, Recommended Standard for Responding to Hazardous Materials Incidents, is accepted throughout the United States in most provinces in Canada. Each level has its own particular applications and criteria for selection.

a. Level A

Figure 3.6 illustrates the features of a Level A suit. Level A offers the highest level of protection available to response personnel for both respiratory and skin protection. Level A should be used in all cases where the

2. Protective Clothing   49

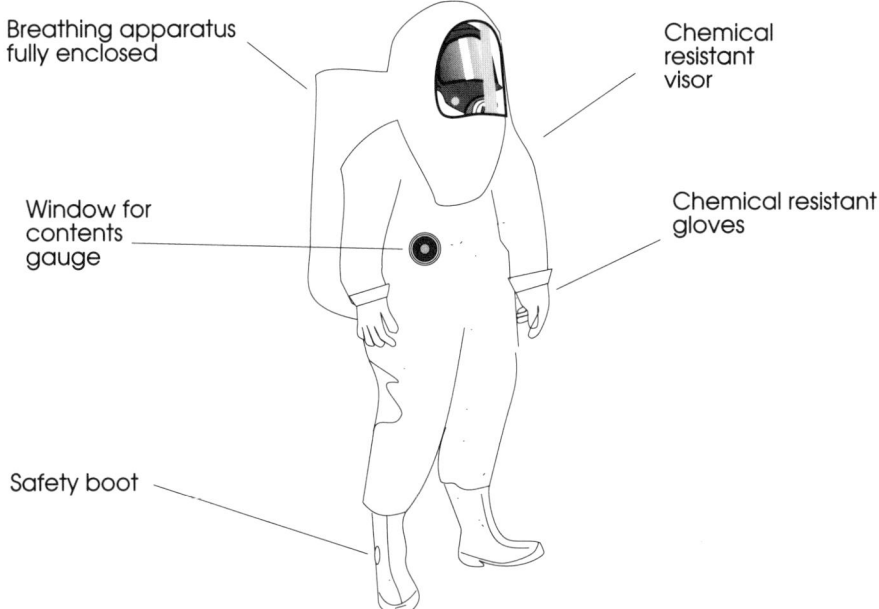

FIGURE 3.6.   Level A Chemical Protective Clothing

material to be handled so indicates (i.e, materials with a dermal risk upon exposure to vapors) and in all cases where the materials involved in an incident have not been clearly identified.

Level A protection constitutes the following:

- Pressure-demand, self-contained breathing apparatus (SCBA) or supplied-air respirator (SAR) with egress unit
- Fully encapsulating, chemical-resistant suit
- Chemical-resistant safety overboots
- Inner chemical-resistant gloves (As a further precaution, during doffing it is recommended that latex inner gloves be worn beneath the first pair of inner gloves.)
- Outer chemical-resistant gloves
- Two-way radio communication (optional)
- Disposable overboot covers and gloves (This eases the decontamination process but is optional dependent on the persistence of materials to be handled.)
- Disposable oversuit (optional)
- Safety helmet (optional)

Before donning a Level A suit, some thought should be given to the clothing to be worn inside it. While there are no hard-and-fast rules, criteria include the following.

- Protection of wearer from suit contact. Prolonged contact between the wearer's skin and the suit material is at best unpleasant and at worst may cause skin abrasions.
- Ambient temperature and sun strength. Generally, the temperature inside a suit will be significantly above ambient. In many cases a light cotton working overall is the ideal undergarment. A major advantage of cotton is its ability to absorb perspiration.
- In extremely cold weather, it may be necessary to use additional, woolen clothing. Care must be taken in making this determination, because even in very low temperatures a wearer inside a suit will heat up rapidly.
- If the material to be handled poses a temperature risk because of its boiling point, then thermal protection against liquid splashes may be necessary. For example, ammonia boils at $-33°C$ ($-27.4°F$). Any contact with the liquid phase, even through chemical protective clothing, may cause freezing or frostbite.

Before using a Level A suit, the following precautions should be taken.

- Inspect the suit for chemical degradation, abrasion, puncture, and seam failure. Normally a visual inspection is sufficient. If there are any reservations concerning the suit's integrity, it should be pressure tested in accordance with the manufacturer's recommended procedures.
- Ensure that the suit is capable of withstanding exposure to the materials to be handled. The manufacturer should supply data that indicates the resistance of the suit to various materials in terms of permeation rates, the consequent breakthrough time, and ability to withstand degradation (the terms *permeation* and *degradation* are discussed later in this chapter). If such data is not available, then the suit should not be worn.
- Assess the degree of mobility restriction in relation to the work to be performed. Level A protective clothing can be very restrictive to movement and often impedes the wearer's vision. In some cases a suit and its material may prove to be so restrictive to mobility as to render a given task unsafe. The problem normally arises with heavier suits, which are designed to provide long duration of use. An alternative may be to sacrifice some duration but gain mobility by selecting a lighter, more maleable material.
- Ensure that the wearer has removed all loose objects, sharp objects, cigarette lighters, and other such items prior to donning the suit. Any hard objects worn inside, protective clothing will increase the likelihood

2. Protective Clothing   51

of suit damage. Of particular concern is the carrying of cigarette lighters. Impingement on the actuator may cause gases to accumulate inside the suit, with a subsequent risk of combustion.
- Consider the amount of time necessary, once the wearer is on supplied air, for suiting up, site approach, site exit, and decontamination prior to suit doffing. If the total work time available becomes impracticable due to the other parameters, then air lines may have to be used in preference to SCBA, or the job assignments to be carried out at Level A will have to broken down into multiple stages.

Table 3.5 compares the advantages and disadvantages of Level A suits.

b. Level B

Figure 3.7 illustrates the features of a Level B suit. Level B gives the highest level of respiratory protection, although a lower level of skin protection is afforded.

Level B protection is constituted by the following criteria:

- Self-contained breathing apparatus (SCBA), or supplied-air respirator (SAR) with egress unit.
- Hooded, chemical-resistant clothing. The NFPA standard allows for splash-type suits or disposable chemical-resistant overalls. Where the potential for liquid contact with contaminants exists, it is recommended that two layers of chemical-resistant clothing be used. The best method identified by this author is the use of a hoodless undersuit and a hooded oversuit, both made of the same material.
- Chemical-resistant inner and outer gloves.
- Chemical-resistant safety boots.
- Although not a requirement, it is recommended that all joints in the array be taped with commercially available 1.5-in. duct tape. This does not guarantee protection against vapors or liquids contacting the skin, but such protection will be enhanced. Key areas to be taped are joints at the neck, wrist, ankles, and face piece.

TABLE 3.5   Advantages and Disadvantages of Level A CPC

Advantages	Disadvantages
Highest level of protection	Bulky and cumbersome
Minimal training requirements	Limited access to breathing apparatus
	Limited duration of use, especially with SCBA
	Initial cost of suit

52 Personal Protection Equipment

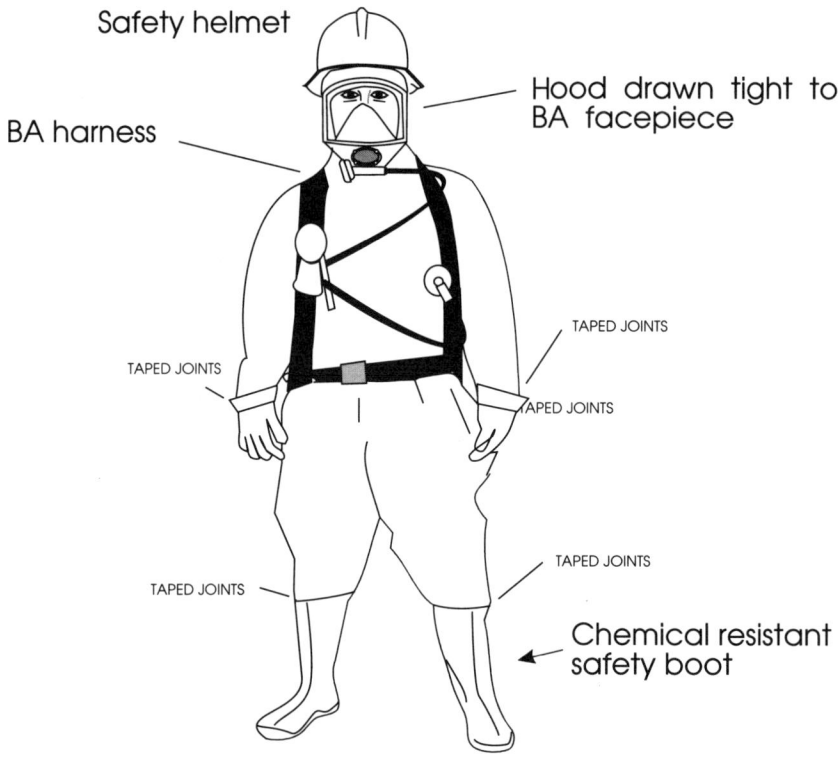

FIGURE 3.7. Level B Chemical Protective Clothing

While Level B gives sufficient respiratory protection for IDLH atmospheres, it should be borne in mind that it does not offer sufficient protection against those materials with a dermal risk, nor does it afford protection to breathing apparatus.

Before donning Level B clothing, the same criteria discussed earlier in this chapter should be assessed, along with a determination as to whether Level A may be a more applicable level of protection.

The donning of Level B protective clothing is fairly involved. The following procedure is recommended.

- Prepare:
  Two suits, one with hood
  Two pairs disposable overboots or integrated bootees
  One pair chemical-resistant safety boots
  Two pairs chemical-resistant gloves

Two rolls duct tape
SCBA or SAR with egress
- Don street clothes, remove shoes, jewelry, cigarette lighters, and sharp objects.
- Don inner suit, inner chemical-resistant gloves, and inner disposable bootees (if not integrated).
- Tape joints.
- Don outer suit, outer chemical-resistant gloves, chemical-resistant safety boots, and face piece.
- Tape all joints and face piece to suit hood.
- Don SCBA or connect air line.

By observing this procedure the best possible protection will be obtained. Table 3.6 compares the advantages and disadvantages of Level B suits.

c. Level C

Level C protection is used when the operator can be fully protected against all hazards by air-purifying respiratory protective equiptment and nonencapsulating garments. Level C may be used only where there is more than 19.5 percent oxygen, the exact nature of all materials is understood, and the IDLH of any substance is not exceeded. Level C protection is constituted by the following;

- Full-face air-purifying respirator (APR) (a half-face-piece unit may be used provided appropriate eye protection is worn) or powered air-purifying respirator (PAPR)
- Chemical-resistant clothing
- Inner and outer chemical-resistant gloves
- Chemical-resistant safety boots

Table 3.7 compares the advantages and disadvantages of Level C protection.

TABLE 3.6 Advantages and Disadvantages of Level B CPC

Advantages	Disadvantages
Low cost	Will not totally protect skin
Lightweight, hence longer duration	Cannot be used for dermally toxic materials
SCBA accessible	Need significant wearer training prior to use
Suitable for atmospheres above IDLH provided thare are no dermal risks	

TABLE 3.7 Advantages and Disadvantages of Level C CPC

Advantages	Disadvantages
Relatively cheap	Only for use in atmospheres that will support life
Easy to use	Atmosphere must be fully characterized and all products catalogued
Lightweight	
Longer duration	

d. Level D
Level D protection represents the minimum standard of protection at any work site. It is for use in the support zone only. Level D protection is constituted by the following;

- Overalls, chemical-resistant safety boots, and hard hat
- Safety glasses or goggles

2B. Selection Criteria

Each level of protection offers advantages and disadvantages in use; generally, the higher the level of protection, the more awkward and cumbersome it is to work in.

The determination of which level of protection is most appropriate is governed primarily by safety, the main objective being to work with the lowest level of protection possible (thereby giving maximum mobility, duration, and comfort) while ensuring adequate protection of the wearer. Figure 3.8 presents a decision logic for protective clothing selection.

Once the level of protection has been determined, other criteria that need to be assessed include the following.

Underclothing. Will clothing selected for wear under the suit offer sufficient protection for the wearer from abrasion of the skin caused by the suit?

Suit compatibility. Is the suit material of sufficient strength for the job to be undertaken? Is the material resistant to the hazardous materials to be, or likely to be, encountered? Are permeation rates and breakdown times acceptable for the circumstances of use? All manufacturers supply data aimed at resolving these questions; an excellent impartial reference is the American Conference of Governmental Industrial Hygienists (ACGIH) *Guidebook for the Selection of Chemical Protective Clothing.*

Decontamination. Can the suit be readily decontaminated after use, or will

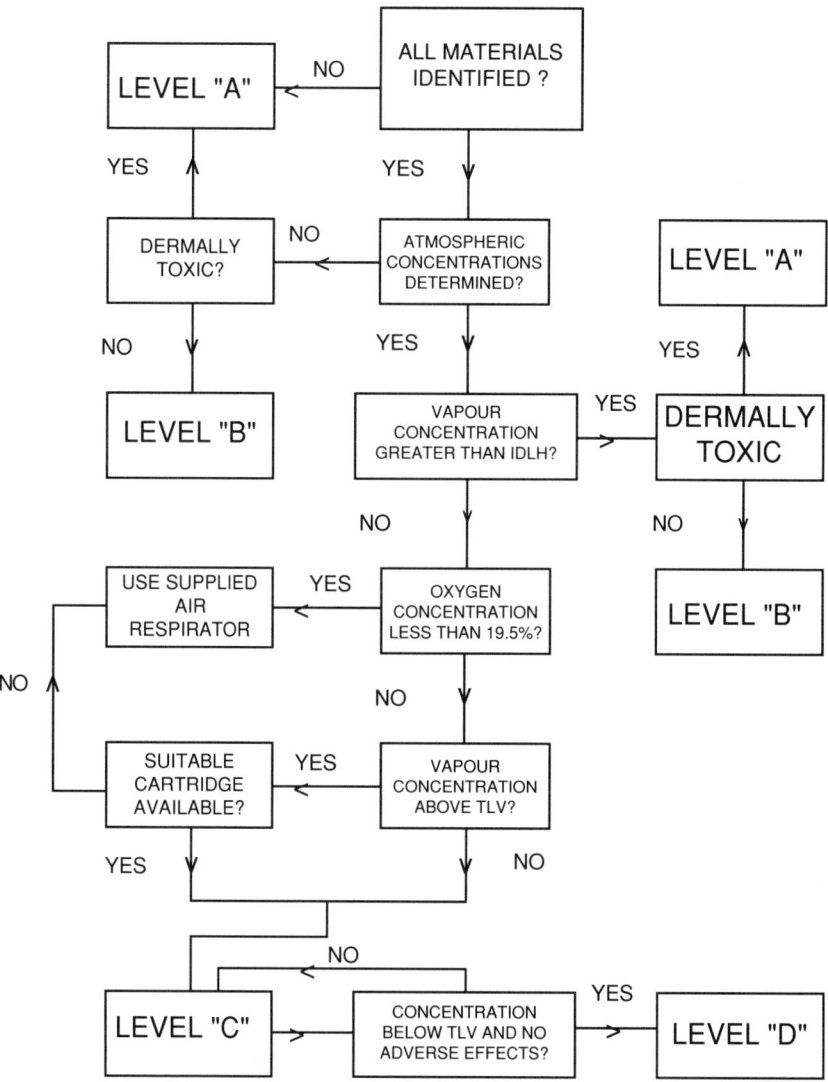

FIGURE 3.8. Decision Logic for Chemical Protective Clothing Selection

it need to be disposed of? In some cases, especially with an expensive reusable suit, it may be desirable to wear a disposable oversuit; then protection is guaranteed by the reusable suit while decontamination is reduced by use of the disposable oversuit.

Equipment compatibility. Not only must a suit be resistant to materials encountered, it must be compatible with other equipment that the wearer will use. For instance, if a supplied-air respirator with egress is to be used for a Level A entry, the suit must be fitted with an appropriate linkage device, commonly known as an IDLH coupling. If a SCBA is to be used, then all controls must be readily accessible by the wearer. Of particular consideration is the ability to access the bypass valve.

Heat. As mentioned previously, use of any level of protective clothing can cause a significant rise in body temperature. In some cases this factor may be so significant that a lighter suit may have to be selected. Heat and the problems associated with it are discussed below.

Temperature effects. In addition to the effects of heat on wearers, there is another consideration involving temperature. Suits themselves are prone to variations in temperature. Not only is their shelf life affected by variation, but significant reduction of mobility may occur in cold conditions; indeed, the suit may become so unmalleable as to be rendered unserviceable. Other problems involve cracking, particularly at the seams and articulation points, loss of integrity, and exhaust valve malfunction. Even if a suit is to be used in normal ambient temperature conditions, consideration should be given to the temperature of materials to which the wearer will be exposed. As mentioned earlier, this is very pertinent when handling gases with low boiling points.

Duration and cost. The duration required of a suit is a strong factor in determining the type of suit to use. Most manufacturers will produce a range of suits for each level of protection, ranging from lightweight, disposable types through to sophisticated, heavy, and costly material combinations that are designed to be reused. Generally, the heavier the material, the longer will be the suit lifetime, resistance to degradation, permeation, abrasion, and liquid exposure. However, it is also true that the heavier suits are considerably more expensive, which may be a significant factor, especially where complete decontamination of a suit is difficult. Cost then is a consideration, albeit not the highest priority.

2C. Durability

Suits are affected by the following factors, which tend to dictate the duration of use:

- Exposure to sunlight
- Variations in temperature
- Degradation
- Permeation
- Mechanical damage

The first two factors are largely uncontrollable; however, care in storage will certainly reduce their effects. Suits, especially, those such as butyl rubber ones, should be stored in a cool, dry place with an even temperature around 15°C (59°F).

Degradation is a chemical reaction in which the molecular structure of the suit material is actually broken down by contact with a contaminant. Typically, the action will cause the suit material to liquefy, swell, blister, or shrink. The degree to which a suit will degrade upon exposure to a contaminant varies with factors such as vapor concentration, liquid exposure, ambient temperature, and humidity. Manufacturers supply data based on such factors.

Permeation is the actual diffusion of a contaminant throughout the suit material at the molecular level. The process may not damage the suit itself and is often not apparent upon visual inspection. If permeation is of a sufficient degree, contaminants may pass right through the suit material. The time taken for this to occur is published and known as the *breakthrough time*. It should be noted that permeation may continue for a considerable period of time after an exposure, particularly if a contaminant has permeated a suit prior to decontamination. As with degradation, manufacturers rate their suits for permeation rates and breakthrough times, for exposures to both liquids and vapors.

A word of caution: Manufacturers' data are supplied for individual products. Mixtures of materials, especially mixtures of chemicals, may cause significantly more rapid degradation and permeation. If, at the planning stage, a possible combination of hazardous materials is recognized, it is advisable to submit material combinations to the suit manufacturer for appropriate testing.

Mechanical damage, such as abrasions, injections, and penetrations, are a major cause of suit failure. Obviously, once a suit surface has been abraded, its ability to withstand permeation is significantly reduced; injection leads to total loss of suit integrity and is remedied by leaving the exclusion zone, decontaminating, and obtaining a new suit. Care must be taken at all times when handling and wearing protective clothing to ensure that mechanical damage is kept to a minimum.

2D. Precautions in Use

For working in protective clothing, the user should note the following.

- Degradation: Are there any physical signs of suit degradation due to exposure? This is particularly important when approaching an incident where not all the materials involved have been identified.

58   Personal Protection Equipment

- Liquids: If a suit wearer comes into contact with liquids, efforts should be made to remove them from the suit surface. At the very least, materials should be brushed off. If it is reasonably practicable, the wearer should go through decontamination procedures. As mentioned before, suit degradation and permeation are significantly accelerated upon exposure to liquids. In dealing with an incident, all possible efforts should be made to keep liquid contact to a minimum.
- Skin irritation: Whenever discomfort or irritation is felt, leave the exclusion zone. In many cases this feeling could be due to perspiration or could be purely psychological. However, one should err on the side of caution: It could be the first indication of suit failure.
- Odor perception: This is also an early indicator of the breakdown of personal protective equipment.
- Physical problems: Any discomfort, breathing difficulties, fatigue, nausea, increased pulse rate, chest pain, etc., is reason for immediately leaving the exclusion zone, decontaminating, and doffing all personal protective equipment. Many of these conditions are heat related and are an early indication of heat stress; in any event, none should be ignored.

2E. Physiological Factors in Suit Use

There are two distinct groups of physiological factors; the first are used in assessing whether a person is fit to use protective clothing at any time, and the second should be assessed immediately before nominating a person to wear protective clothing.

In determining if a person should be preselected for working in protective clothing, the following criteria should be assessed.

Fitness. Working in chemical protective clothing and breathing apparatus (particularly self-contained breathing apparatus) is physically demanding. For this reason, any condition that inhibits normal working ability should negate selection, including any debilitating disease.

Ability to climatize. Not everyone is capable of working in chemical protective clothing. For some the psychological stress is so great that they feel claustrophobic upon donning protective clothing. This condition can often be identified during training, and anyone exhibiting signs of claustrophobia should not be allowed to wear chemical protective clothing. Suit claustrophobia can be succinct or readily apparent in its manifestation: The wearer has only one goal—to get out of the suit. In the succinct case, the wearer consumes breathing air at an excessive rate, often at twice the rate of consumption when not wearing protective clothing.

Size. Generally, chemical protective clothing is manufactured in two or three standard sizes; most often these are medium, large, and extra large. Personnel outside standard fit ranges should not be selected to wear chemical protective clothing unless they can be supplied with a tailored suit.

Obesity. Obese people will suffer heat stress sooner and to a greater degree. Additional complications involve general physical stress and suit sizing.

Immediately prior to working in protective clothing a wearer should be assessed for;

- Consumption of alcohol. In addition to the motor problems associated with alcohol consumption, wearers will be more susceptible to heat stress conditions.
- Medication or controlled substances. Any person taking medication should not wear chemical protective clothing.
- Body temperature. Elevated body temperature means that the wearer is starting out from a higher base temperature than normal. Such elevation may be due either to early signs of heat stress or the presence of infection/disease. All such conditions preclude the wearing of chemical protective clothing.
- Sunburn. This may cause elevated temperatures, dizziness, nausea, and increased susceptibility to heat stress. Even mild sunburn may preclude the wearing of chemical protective clothing.
- Adequate rest. If a person is not adequately rested prior to wearing chemical protective clothing, fatigue and other such conditions may set in much earlier than in other personnel. Additionally, air consumption is likely to be higher.

2F. Heat Stress

Throughout this chapter references have been made to the problems associated with heat. The problems vary in intensity and depend on temperature, duration, suit type, and the physiology of the individual. Table 3.8 illustrates the progressively associated problems together with their symptoms and basic treatment.

The best way to avoid heat stress is by monitoring personnel and taking adequate precautions. While there are several sources of discussion on the exact values that indicate heat stress, a general rule is that if the pulse

TABLE 3.8 Signs and Symptoms of Heat Stress

Heat Rash	Heat Cramps	Exhaustion	Stroke
Angry red rash	Muscle spasm	Pale, cool, moist skin	Hot dry skin
Wash & talc	Pains in hands & feet	Heavy sweating	Nausea
	Rest and increase fluid/salt intake	Dizziness & nausea	Strong, rapid pulse
		Fainting	CAN BE FATAL
		Rest and increase fluid/salt intake	GET MEDICAL HELP

Source: US EPA. 1984: *Standard Operating Safety Guides.* Office of Emergency & Remedial Response. Hazardous Response Source Division, Edison. N.J.

rate of an individual exceeds 110 beats per minute or if the oral temperature exceeds 37.5°C (99.6°F), shorten the length of time that a person is working in protective equipment.

In any event, if body temperature exceeds 37.8°C (100°F), do not allow the individual to wear protective clothing.

Preventative measures that will help reduce heat stress include:

- Selection of the lightest, most permeable protective clothing that will afford safe protection
- Reduction of duration in protective equipment
- Frequent rest periods
- Air-conditioned shelter for rest spells
- Drinking plenty of water, supplemented by glucose
- Suit cooling devices (ice packs, air line cooling)
- Avoiding alcohol and medication
- Keeping fit

3. DECONTAMINATION PROCEDURES

Effective decontamination is an issue that is often overlooked at incident sites; indeed, in the early critical stages it is often regarded as a luxury rather than a necessity. Decontamination of all articles leaving the exclusion zone is essential for the prevention of contaminant spread. Not only must entry team members themselves be protected against exposure but also other team members and the environment at large.

As part of the planning process, before the occurrence of an incident, a formal decontamination plan should have been developed and included in the contingency plan. A decontamination plan should also be written for use at sites without contingency plans. Entry into the exclusion zone should not be permitted without a decontamination procedure in place. The decontamination plan should address the following:

- Layout of decontamination site
- Number of personnel required for decontamination
- Roles and functions of the decontamination team and its supervisors
- Methods to be used in decontaminating
- Protection levels for the decontamination team
- Water treatment and disposal methods
- Contaminant treatment and disposal methods
- Emergency decontamination procedures
- Post-decontamination procedures

Situations may arise where a team has to respond with no knowledge of the products involved. In this extreme circumstance, a decision must be made concerning the advisability of decontamination. It is possible that the process of decontamination may cause a chemical reaction or the evolution of toxic materials, with subsequent risk of exposure to personnel. If there is no decontamination procedure, then all possible steps must be taken to avoid contact with any liquids or vapors to prevent the spread of contaminants into the support zone.

3A. Site Setup

The decontamination process takes place in the contamination reduction corridor, which forms part of the contamination reduction zone. Figure 3.9 shows the relationship of these zones. Note the indication of the wind direction. Safety of personnel in the support zone should also be considered in the event of a change in wind direction.

While the exact site layout will vary depending on the decontamination protocol to be observed, all sites will include some or all of the following:

- Water sprays
- Spray recovery sheets/baths
- Water recovery tanks (portatanks)
- Detergent buckets
- Mops
- Brooms
- Wipe rags

Figure 3.10 illustrates a typical decontamination station layout for a five-stage process.

3B. General Decontamination Methods

There are two essential methods of decontamination: physical and chemical. Physical methods include the following:

Personal Protection Equipment

Decontamination Station Layout

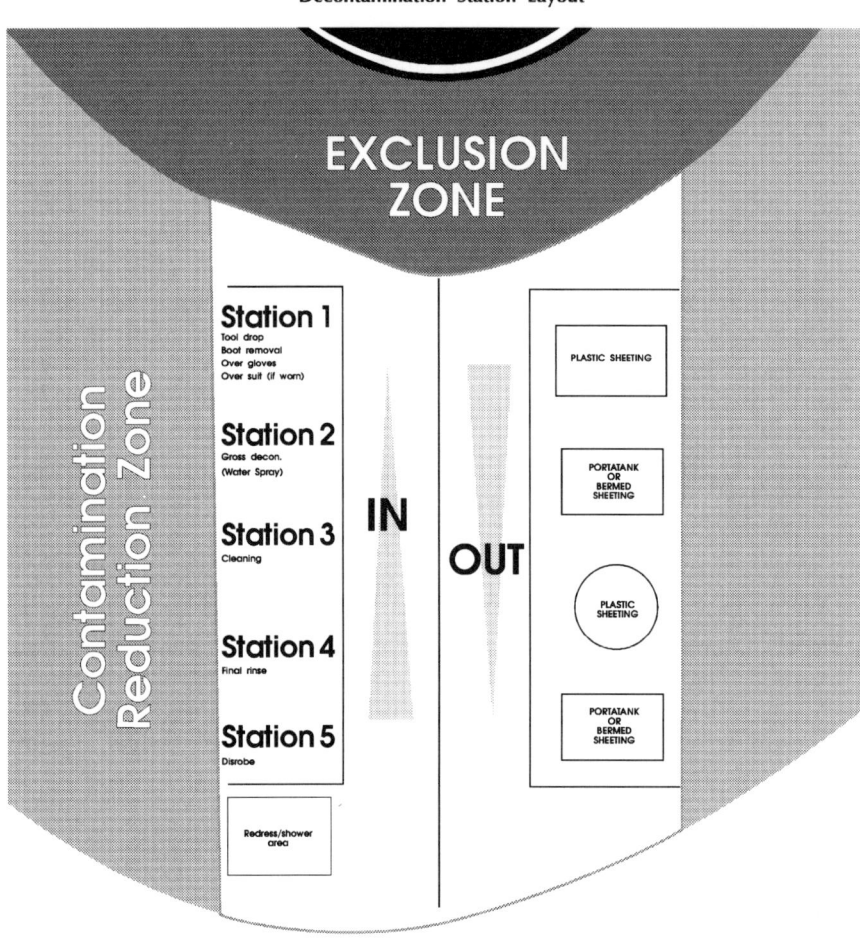

FIGURE 3.9. Incident Control Zones

- Scraping, to physically dislodge tenacious materials that are in direct contact with equipment and apparel.
- Brushing, similar to above.
- Rinsing, normally with low-pressure water (often from fire hoses). This is the most often-used method for gross decontamination.
- Vacuuming. This method is used primarily for removal of particulate contaminants. Recovered particles tend to become airborne and hence

3. Decontamination Procedures   63

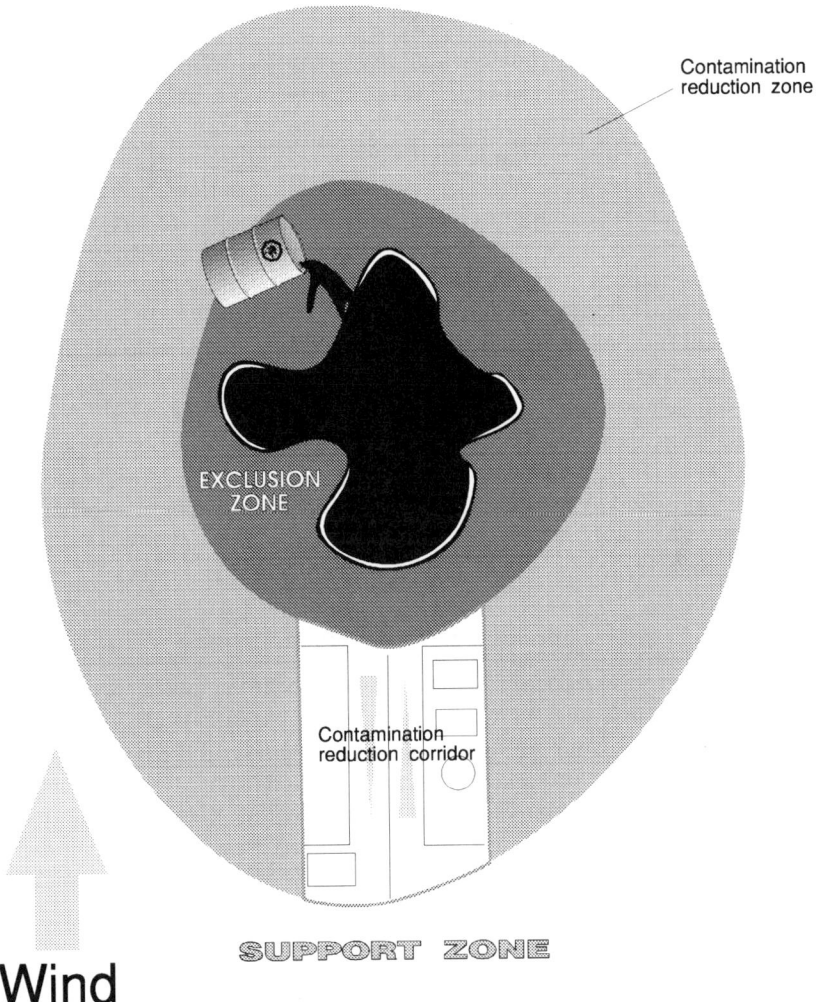

FIGURE 3.10.   Decontamination Station Layout

a source of contamination. For this reason it is important that a vacuum cleaner fitted with a HEPA filter be used.
- Heating. Some contaminants may need heating prior to removal. An example is the use of hot detergent water rather than cold for the removal of diesel oil.
- Cooling. On occasion it may serve better to cool a contaminant to the point where it is brittle enough to be broken and removed.

Chemical methods include the following:

- Dissolving, by the introduction of an intermediate substance during the decontamination process (e.g., using kerosene as an intermediate for heavy fuel oil decontamination).
- Neutralizing. This is often used for corrosive substances; when an acid is being handled, a base can be used for neutralization, and vice versa.
- Surfactants. This technique is used as an enhancement to physical cleaning and often as a follow-up to dissolving. Common surfactants used include trisodium phosphate (TSP), industrial-grade detergents, and common household cleaners.
- Solidification. Several patent gelling products are available on the market. Their application causes the contaminant to solidify, hence making removal easier.
- Sterilization, used mainly for etiological contamination. If nothing better is available, rinsing with household bleach is better than no decontamination at all. However, most etiologic substances are very difficult to decontaminate effectively, and often isolation and disposal is the preferred method.

3C. Specific Decontamination Methods

The exact method of decontamination for a specific material will depend on many variables, such as temperature, degree of exposure, presence of other materials, resources available, and manufacturers' recommendations. The guidelines in Table 3.9 suggest general principles in decontamination. Even when contamination is not suspected, it is good practice at least to rinse down all wearers and equipment as they leave the controlled areas. Table 3.9 is based on the following five perceived levels of contamination:

Level A. Basic cleanup for use in any situation where personal protective equipment has been used. Essentially applies to vapor exposure or where no exposure at all is suspected.

Level B. Contact with any substances that cause a dermally toxic, corrosive, or reactive risk. Also used for water-soluble contaminants and insoluble contaminants with low viscosities (i.e., gasoline).

Level C. Possible contact with unidentified contaminants or with contaminants that are water insoluble (high viscosity), highly reactive, toxic by inhalation, or that have acute systemic toxic effects (i.e., cyanides).

Level D. Decontamination without water. Used primarily for water-reactive contaminants. Normally supplemented by Level A, B, or C decontamination procedures.

TABLE 3.9 Specific Methods of Decontamination

Decontamination Level	Procedure		Decontamination Team Protection
A	1	Rinse with water. Brush off any large deposits	Level C protective clothing and equipment
	2	Wash with surfactants	
	3	Re-rinse and then exit decontamination area	
B	1	Flush with water. Brush off any large deposits	Level B protective equipment (disposable clothing recommended)
	2	Wipe with intermediate agent/neutralizer	
	3	Rinse	
	4	Wash down with surfactant	
	5	Re-rinse and then exit decontamination area	
C	1	Follow steps 1 to 5 as above	Level A protective clothing, depending on nature of contaminant
	2	Drum or bag all materials and protective clothing for further decontamination and testing.	
	3	All decontaminated personnel must wash and shower immediately	
	4	Inner clothing should be bagged or drummed for further testing	
D	1	Vacuum surface into industrial cleaner fitted with filtration system	Level B protective equipment (disposable clothing recommended)
	2	Dust with fullers earth or talc and re-vacuum	
	3	Follow steps 3 thro' 5 as in B above	
	4	Drum or bag all materials and protective clothing for further testing	
E	1	Flush with water	Level A protective clothing and equipment
	2	Spray with 10% chlorine bleach solution	
	3	Flush with water	
	4	Do not attempt approach before consulting CANUTEC or CHEMTREC	

General Notes:
1. All wash water must be recovered, treated and/or disposed.
2. Equipment should be retained inside contamination reduction zone or exclusion zone until final decontamination procedures are implemented.

Level E. Decontamination for exposure to etiologic agents. This process can be extremely difficult, and disposal may be a better option. Before attempting to deal with etiologic agents, contact CANUTEC, (613) 996-6666, or CHEMTREC, (800) 424-9300, for advice on all phases of incident management.

In addition to the decontamination procedures outlined previously, all personnel should shower before leaving the incident site, or as soon as practicably possible thereafter. All other personal hygiene should be constantly emphasized to reduce the risk of exposure to contaminants. This is particularly important prior to the consumption of food.

3D. Emergency Decontamination

Emergency decontamination is reserved for circumstances such as victim decontamination, injured response personnel, failed personal protective equipment, and other such occurrences. Generally, the preferred method is to decontaminate those areas on the victim or person where contact with others is necessary and then to cover all other parts of the person and protective apparel with plastic sheeting or similar protection.

In many cases emergency decontamination at an incident site will be carried out under extremely stressful circumstances. Despite this, it is important that sufficient decontamination take place to ensure that the life and health of other parties is not threatened. Other personnel at a response site may be affected by contaminants. Due consideration should also be given to paramedics, hospital personnel, and others. When a contaminated victim is evacuated, the exact nature of the contaminants must be clearly indicated to those providing care. Emergency decontamination should be included as a part of the contingency plan and its decontamination plan.

3E. Post-Incident Procedures

When a person leaves the exclusion zone, perhaps to change air bottles or for consultation, it is common for that person to be decontaminated only to a level sufficient to allow temporary doffing. Once the person is ready to reenter the exclusion zone, he or she will probably dress again in existing personal protective equipment and collect contaminated tools and instruments from the contamination reduction zone upon reentry.

When no further entry is contemplated, or if the level of protection/decontamination procedures is to be downgraded, then a full decontamination of personal protective equipment and tools will be necessary.

In all cases, before equipment is finally stored, it should be thoroughly cleaned and then inspected to ensure that decontamination has been effective. Methods include wipe testing and analysis, exposure to ultraviolet light, and visual solution or permeation testing.

For some substances it may never be possible to effect total decontamination; in such conditions the only option left may be disposal. For example, after PCB-related incidents, whole backhoe assemblies have required disposal.

3F. Prevention

The most effective method of decontamination is the prevention of contamination. Several procedures will aid in achieving this goal:

- Avoid contact. Do not stand in pools of liquid; avoid walking in areas where a material is leaking; use grapples, hooks, etc., to touch contaminated equipment and apparatus.
- Bag. Wherever practicable, tools, monitoring instruments, and other items that must be taken into the exclusion zone should be bagged. While this may not totally negate the need for decontamination, it should reduce it to a minimum.
- Encapsulte. If an object is contaminated and it is likely to be handled or brushed against, it should be encapsulated (in plastic sheeting, for example).
- Use overclothing. For example, if wearing a reusable suit while dealing with a phenol spill, the use of a disposable oversuit will make decontamination of the considerably more expensive, reusable suit much easier.

4

Incident Risks and Safety

This chapter reviews the dangerous properties of hazardous materials, potential dangers at incident sites, and ideas on how to approach an incident safely. A full discussion of the characteristics of hazardous materials would comprise a book in itself; indeed, a strongly recommended volume is *The Common Sense Approach to Hazardous Materials,* by Frank Fire. This chapter is divided into nine sections:

1. Flammable risk of liquids
2. Flammable risk of gases
3. Corrosivity
4. Reactivity
5. Physiological effects
6. Oxygen deficiency
7. Hazard indication
8. Incident risks
9. Site assessment and safe approach

The goal of this chapter is to illustrate briefly the properties of hazardous materials and show methods by which they may be recognized.

1. FLAMMABILITY OF LIQUIDS

Many liquid chemicals and refined products are flammable. When liquids burn, it is not the liquid itself that is burning but the vapors liberated from the liquid at or near its surface. In order for a liquid to burn, there must be sufficient vapor at the surface to support combustion. Such a vapor

1. Flammability of Liquids   69

concentration must also occur in the presence of adequate oxygen and a heat source of sufficient energy to ignite the mixture.

The amount of vapor needed to support combustion at a liquid surface will depend primarily on the volatility of the liquid, i.e., the amount of vapor emitted at a given temperature. At relatively low temperatures the liquid may not give off sufficient vapor for combustion to occur. Raising the temperature of a flammable liquid will eventually cause sufficient vapor to be emitted for support of combustion.

In characterizing flammable liquids, a simple extension of the above process is performed. A liquid is heated in an apparatus similar to that shown in Figure 4.1. During heating, a source of ignition is constantly introduced to the liquid surface. Eventually a temperature is reached at which "flashover" occurs at the liquid surface. This temperature is known as the *flash point* and is the lowest temperature at which a liquid releases sufficient vapor to form an ignitable mixture with air near the surface of the liquid.

The flash point should not be confused with boiling point, which is the temperature at which the vapor pressure of a liquid equals atmospheric

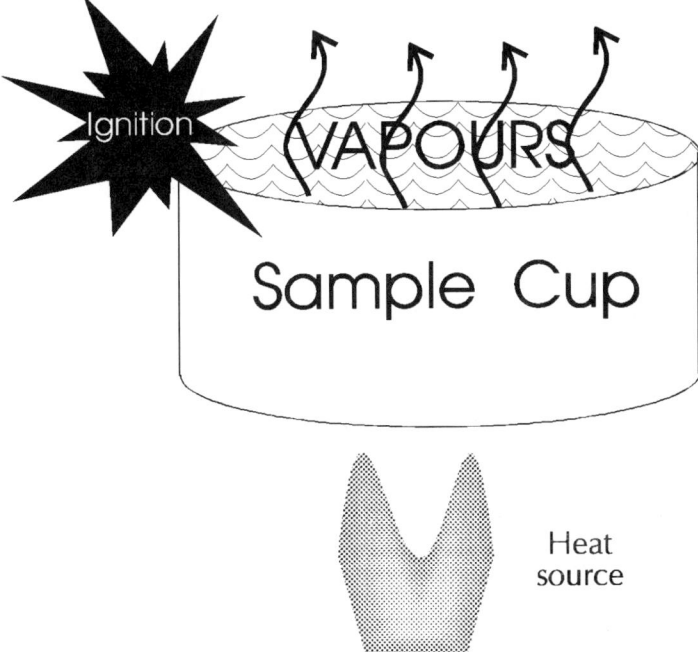

FIGURE 4.1.   Open-Cup Flash-Point Test

TABLE 4.1   Interrelation of Terms and Values Used to Express Flammability

Material	Melting Point	Flash Point	Boiling Point	Auto Ignition Temperature
Butane	−138 C	−60 C	−1 C	360 C
Hydrogen	−259 C	<−50 C	−253 C	500 C
Furfural	−37 C	60 C	162 C	316 C
Butyl-Alcohol	−89 C	35 C	83 C	350 C
Toluene	−95 C	13 C	111 C	480 C

pressure. The boiling point of a liquid will normally be higher than its flash point.

As will be discussed later, flammable liquids are generally taken to be those with a flash point of less than 61°C (141.8°F).

Another term used in defining the potential threat of a flammable liquid is the *auto-ignition temperature* (AIT). This is the temperature at which a liquid will ignite without an external source of ignition; on occasion it is referred to as the *fire point*. Table 4.1 shows the interrelation of the various terms used in defining the properties of flammable liquids.

2. FLAMMABILITY OF GASES

When vapors are liberated from contact with their liquid phases, they are known as gases. The distinction between a liquid and a gas is determined primarily by the pressure exerted by a material at a given (standard) temperature. For instance, one standard states that any material that exerts a pressure greater than 275 kPa at 21.1°C (69.9 °F) will be classed as a gas.

Many gases are flammable in nature and, once ignited, may be completely consumed in a sustained combustion. For an atmosphere to be flammable, two essential elements must be present:

- Oxygen, in sufficient quantity to support combustion
- Fuel, in the form of flammable vapors at a concentration that will support combustion

At incident sites, on open ground, there will normally be sufficient oxygen to support combustion; the normal concentration of oxygen in air is 20.8 percent. The concentration of flammable vapors depends on several criteria, including the following.

- The phase of the materials (gas/liquid). Some flammable materials may exist either as a liquid or a gas, depending on ambient temperature. An

example is butane, which boils at −1°C (30.2°F); at temperatures below this value, the propagation rate of vapor will be reduced significantly. Above its boiling point, butane exists as a gas. Obviously, the more heat is applied to liquid butane, the greater will be the volume of gas generated.
- Ambient temperature. At a higher ambient temperature a greater volume of vapor will be generated. Again considering butane, if a quantity of liquid butane is spilled on the ground at a temperature above −1°C (30.2°F), it will boil. However, the latent heat required to vaporize the liquid butane will actually cause the liquid to cool to its boiling point, where it will stabilize. Further liberation of vapor would then depend on the heat supplied from the environment surrounding the butane. The higher the ambient temperature, the greater will be the volume of gas liberated.
- Wind velocity. This will affect the concentration of gas, normally measured as a percentage by volume. The higher the wind velocity, the greater will be the dispersion and hence the lower will be the concentration at a given point. The converse is also true.
- Vapor density. Many flammable gases are heavier than air and tend to accumulate at lower levels, creating a greater risk of flammability in low-lying areas. For gases that are lighter than air (e.g., hydrogen), vertical dispersion is rapid; unless the gas is trapped, concentrations are unlikely to cause sustained combustion.

Given an oxygen concentration of 20.8 percent, there are two limits of concentration that are likely to support combustion:

- The *lower flammable limit* is the lowest concentration of vapor at which sustained combustion occurs.
- The *upper flammable limit* is the highest concentration of vapor at which sustained combustion continues.

Some jurisdictions use the terms *lower explosive limit* (LEL) and *upper explosive limit* (UEL), or *lower inflammability limit* (LIL) and *upper inflammability limit* (UIL). All three sets of terms are synonymous. LFL and UFL are most correct technically, but, LEL and UEL are most commonly used.

Explosion in this case is a term of definition and relates to the inability of a container to withstand the pressure exerted by expanding gases during sustained combustion.

The values (which are expressed as percent vapor by volume) vary considerably depending on the substance. Exact values are determined by

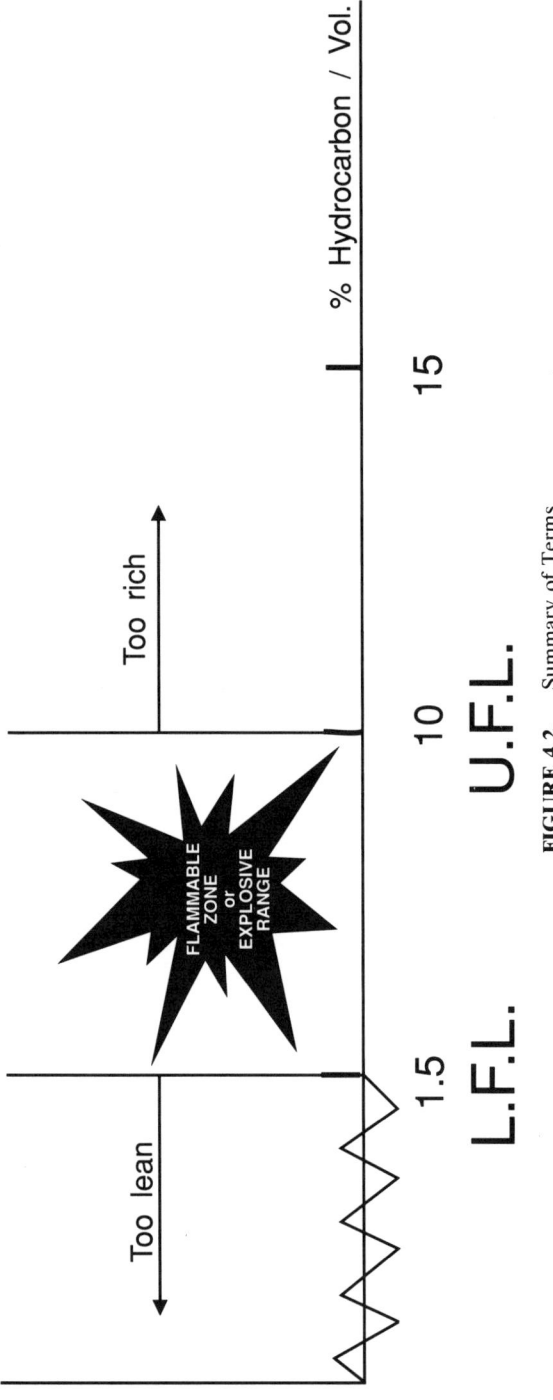

FIGURE 4.2. Summary of Terms

laboratory test methods and are promulgated in such resources as material safety data sheets and emergency response guides. Typical values range from 2 to 20 percent vapor by volume, but there are extreme cases (ethylene oxide, for example, has a LEL of 3 percent and an UEL of 100 percent).

Values quoted in references are always for a concentration of 20.8 percent oxygen (normal air) by volume; distortion of the values will arise in oxygen-enriched or oxygen-depleted atmospheres. Taking the upper and lower flammable limits into consideration, we can define a zone in which there is a risk of sustained combustion. Consider a substance with a LFL of 1.5 percent and an UFL of 10 percent. The range between these two values is known as the *explosive zone* or *flammable range*.

Concentrations below the LFL will not support combustion as there is insufficient fuel; these concentrations are known as *too lean*. Concentrations above the UEL will not support combustion as there is too much fuel; these concentrations are known as *too rich*. Figure 4.2 summarizes these terms.

As mentioned previously, the concentration of oxygen will affect the flammable zone. Enriched atmospheres (those with a concentration greater than 25 percent) will cause extension of the UFL and LFL. Similarly,

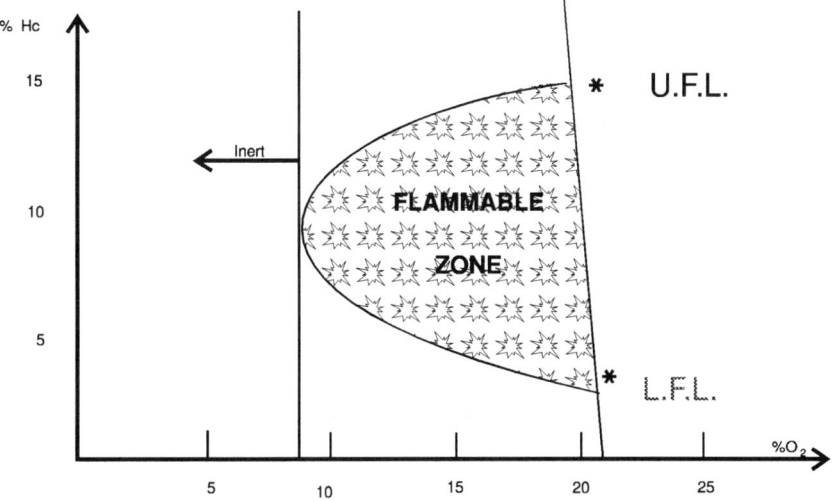

FIGURE 4.3. Effects of Oxygen Concentration

depleted atmospheres (those with a concentration less than 19.5 percent) will cause depression of the values. A point is reached where, regardless of the concentration of flammable vapor present, combustion will not occur. This situation is known as an *inert atmosphere* and is usually quoted at 8 percent oxygen by volume. Figure 4.3 summarizes these terms.

Flammable liquids and gases in containers present an additional risk of explosion in the presence of fire, a condition known as a boiling-liquid, expanding-vapor explosion (BLEVE). An example of this type of explosion is shown in the sequence of diagrams in Figures 4.4 through 4.10.

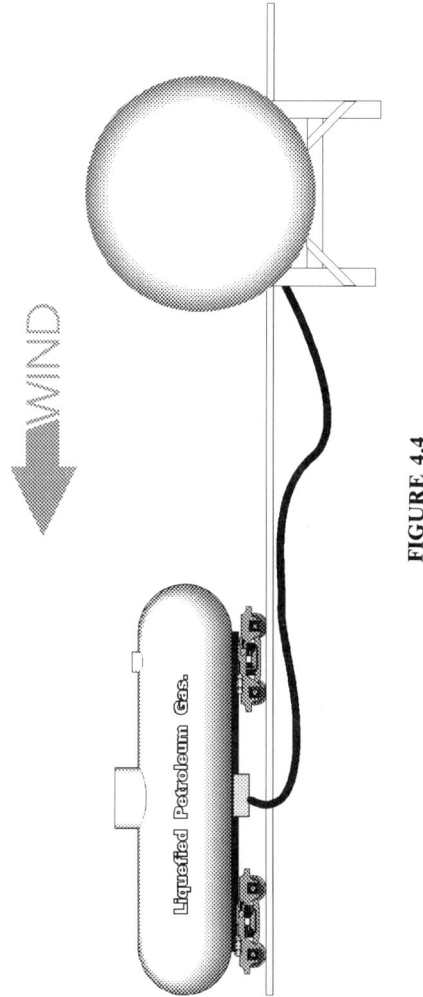

FIGURE 4.4

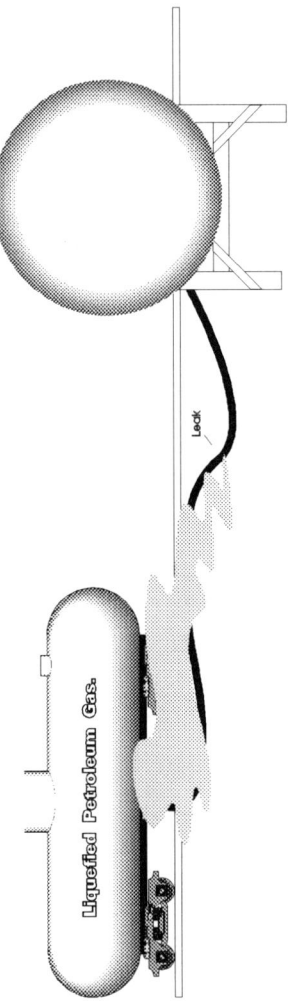

FIGURE 4.5

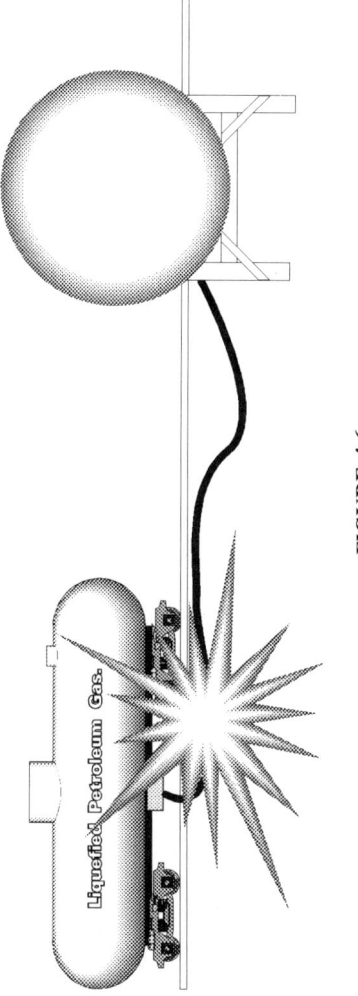

FIGURE 4.6

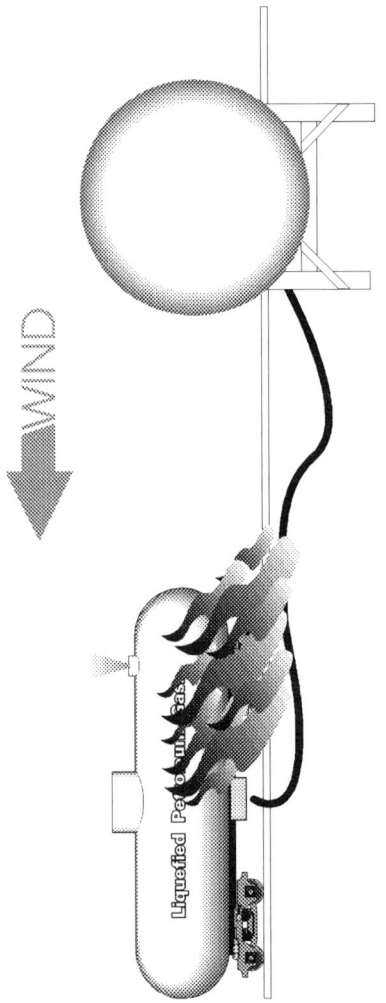

FIGURE 4.7

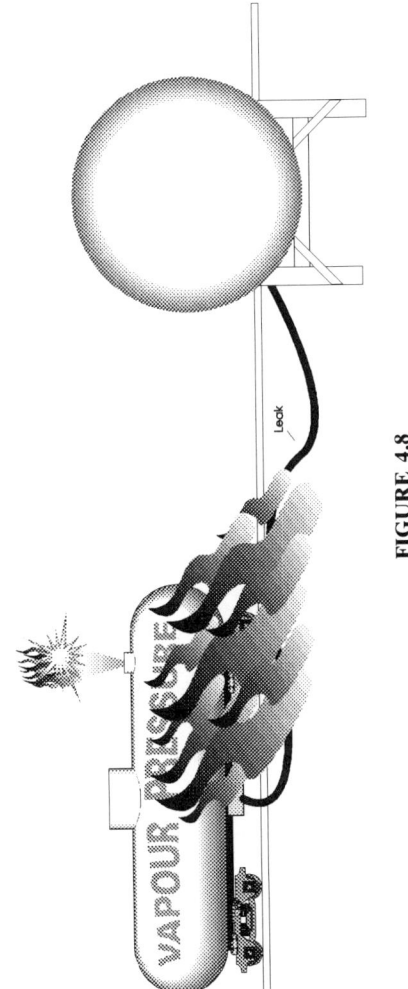

FIGURE 4.8

FIGURE 4.9

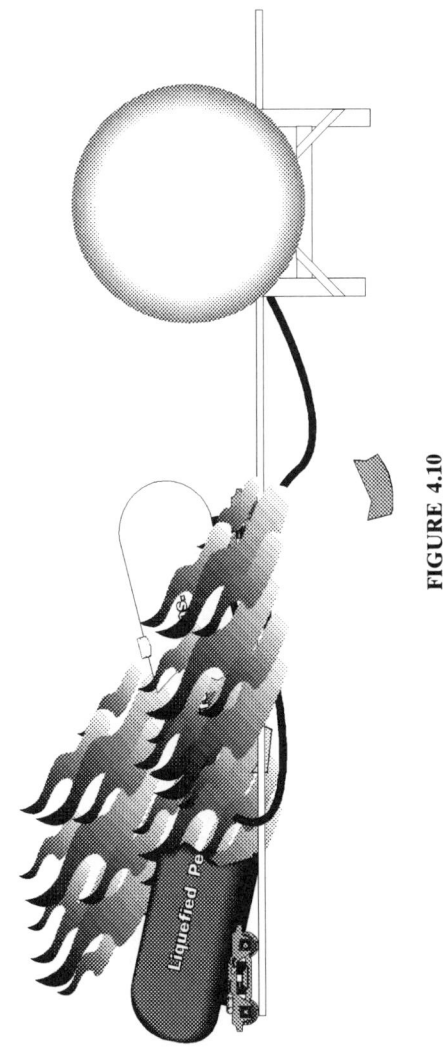

FIGURE 4.10

A propane tank is transferring product to a fixed tank at an industrial facility (Figure 4.4).

During transfer operations a leak develops in the hose. Propane vapors are driven toward the car (Figure 4.5).

The propane contacts a source of ignition, with resulting flashback and fire (Figure 4.6).

As the liquid inside the tank is heated by the external flame, vapor pressure increases inside the tank. Eventually the pressure safety valve lifts. Heat in the lower regions of the tank is constantly dissipated into the liquid propane (Figure 4.7).

The venting gases ignite. Flame impingement and radiant heat cause the tank shell structure, that is in contact with the vapor space, to heat and weaken. Pressure within the car continues to increase (Figure 4.8).

Continued weakening of the storage tank due to heating and rising pressure cause the tank material to tear. This allows rapid depressurization of the liquid propane and the evolution of large quantities of vapor, which instantly ignite (Figure 4.9).

The force generated by the burning gases causes further tearing and then propulsion of what is now the two tank halves (Figure 4.10).

The whole process described in this sequence can occur in less than 10 minutes. During the explosion, components can be propelled considerable distances at extremely high velocities; whole tankcars have been driven hundreds of meters at incidents. The BLEVE is normally accompanied by a large vertical fireball and a horizontal ground flash.

It should be borne in mind that a BLEVE can occur anywhere that flammable gases or liquids are inside a container. Wherever risk of BLEVE is present, evacuation must be an early consideration. Unless large volumes of water can be brought to bear on all the containers, it is highly probable that they will rupture.

3. CORROSIVITY

Many chemicals used in industry are corrosives. A *corrosive* is any material, liquid or solid, that may or has been known to cause damage to human skin or other materials. This definition is extremely broad and varies by legislative jurisdiction. For example, the U.N. definition differs from the U.S. Department of Transportation definition, which differs from the Canadian Workplace Hazardous Materials Information System definition.

Corrosives are divided into two main categories, acids and alkalines (also known as bases). The degree of acidity or alkalinity is measured on a scale of 1 to 14, known as pH. A pH of 7 indicates a neutral material; values above 7 indicate an alkaline material and values below 7, an acidic material. As will be discussed later, pH is used as a criterion for inclusion as a corrosive in hazard classification.

Strong corrosives cause secondary problems beyond those of material corrosion. These effects include the liberation of corrosive/toxic vapors, generation of heat, combustion of other materials, liberation of oxygen, and destabilization of other materials.

A good example of these secondary effects is the action of sulfuric acid. If concentrated sulfuric acid is mixed with a small volume of water, it will liberate large amounts of heat with subsequent fuming and corrosive/toxic gases. If the same material is spilled on an organic substance, it may cause ignition. Spill sulfuric acid on a metal, and it may liberate hydrogen and heat, which could in turn cause fire and explosion. Many other strong corrosives will cause similar effects.

4. REACTIVITY

Many chemicals are produced with the fundamental objective of inducing a reaction when mixed with other chemicals or placed in the presence of a catalyst. Chemicals may react with any or all of the following.

- Other chemicals. These reactions may cause liberation of heat, fuming, or the generation of toxic gases. In extreme circumstances the mixing of two stable chemicals may produce a third highly reactive and explosive material. For example, if diesel oil and ammonium nitrate fertilizer are allowed to mix, they will probably explode.
- Air. Some materials will react with the oxygen and/or moisture present in air. These chemicals are stored and transported in sealed containers, often under a blanket of inert material. If released in a spill and exposed to air, these materials will react violently.
- Water. As in the case of sulfuric acid, many other corrosives will react violently with water, particularly those in strong formulations. Other substances that react with water include some metals and oxidizers.
- Common substances. In a spill situation, a chemical may react with many common substances in the environment, for example, soils, concrete, and wood.
- Themselves. Some materials are *self-reactive*. One example of this is the process of polymerization, where the molecules of a chemical realign into long chains with the subsequent release of heat. Two examples of substances that are susceptible to this process are styrene and ethylene.

Self-reaction is induced by interferences such as heat, light, shock, contamination, or ageing. Most materials of this nature are assessed as having a *self-accelerating decompostion temperature* (SADT). This is the temperature at which they will start to decompose and liberate heat. The more these materials decompose, the hotter they become; as the temperature rises, decomposition accelerates and a chain reaction occurs. Many such materials will self-combust and/or explode once excited.

5. TOXIC RISK

Up to this point primary consideration has been given to the results of exposure to larger quantities of materials. We have reviewed issues such as flammability and explosion. Many hazardous materials can cause harmful effects at concentrations significantly below those that we considered in the previous sections. For example, benzene will create a flammable vapor when the concentration in air reaches 1.3 percent. The first step in addressing an incident involving benzene is to assess the risk of flammability. However, benzene may also induce cancer of the blood after long exposure to very low concentrations. The current "safe" level for benzene is 1 ppm (OSHA 1910.1028). The lower explosive limit of 1.3 percent vapor by volume corresponds to 13,000 ppm, significantly above the "safe" level. As will be discussed in Chapter 5, assessments of explosivity normally give a best accuracy of approximately 2 percent of the lower explosive limit, or 260 ppm for benzene.

Many other hazardous materials also present significant health risks at very low concentrations. This section presents an overview of the risks, how they may be equated to currently documented values, and how some exposures may be detected. This section is not designed to be an exhaustive explanation of toxicology, merely to offer guidance. Suggested for further reading is *Industrial Toxicology, Safety and Health Applications in the Workplace,* by Phillip L. Williams and James L. Burson.

5A. Effects of Toxins

A *toxin* is any substance that upon exposure may cause harm to humans without mechanical damage. Toxic actions are divided into two broad categories, acute and chronic. Acute affects are those that occur rapidly, are readily observable, and may be reversed. An example is exposure to a high concentration of naptha fumes; the victim becomes disoriented and may stumble, fall, and become unconscious. Removal to fresh air, however, results in rapid recovery.

Chronic effects are a result of cumulative and/or repeated low-level exposures over longer periods of time. They are not readily observable and may be irreversible. For example, as noted above, repeated exposure to benzene at low concentrations may eventually induce cancer of the blood. It should be noted that acute affects may become chronic.

The criteria for deciding what makes a toxin a poison is essentially the dose rate, i.e., the rate at which the body absorbs the toxin. For example, by taking two tablets for a headache one day the body is utilizing a potential toxin to achieve an effect, at a dose rate that the body can detoxify. Consumption of many tablets at one time—an overdose—will cause an acute response. An overdose quickly exhausts the body's ability to detoxify the substance, and organ damage may result. Chronic poisoning can result from prolonged ingestion of the correct dose.

Whether or not the body receives a sufficient dose of a toxin to cause adverse health effects is greatly dependent on the type of the toxin and the route of entry into the body. There are four routes of entry into the body.

The first, inhalation, is the one most often considered. This is rightly so, as the lungs have a large surface area that readily absorbs toxins. The second route is dermal, i.e., absorbtion through the skin. Not all toxins can be absorbed by this route; however, some may be absorbed in either the gaseous or liquid phase. If a toxin can be absorbed through the skin, it is given the designation TLV (skin). Typical examples of such substances include phenol and cresol. It should be noted that this designation signifies only that the contaminant will penetrate skin and does not indicate possible skin irritation, corrosive effects, or production of dermatitis. The third route is ingestion. While ingestion of a pure product is highly unlikely, ineffective decontamination or personal hygiene can cause toxins to be ingested when food is handled. Once consumed, many toxins will produce significant toxic effects. The fourth route, puncture or injection, is most commonly used in laboratory testing for determining the potency and effect of toxins. However, should a person suffer any form of wound at a site, and the object causing the laceration or puncture is contaminated, then toxins may be introduced directly into the bloodstream. This route then produces marked effects.

The route of entry depends on several factors. The primary consideration is whether the toxin is a solid, a liquid, or a gas. Generally, gases may enter the body by all routes, characteristically by inhalation; liquids may be absorbed dermally or by ingestion; and solids are absorbed primarily by ingestion. Airborne particulate liquids and solids may also be absorbed in the lungs.

Dose rates aside, once a toxin impacts the body, its exact effect will depend on the individual's physiology. Pertinent factors include:

- Condition of the respiratory system
- Condition of the cardiovascular system

- Condition of the kidneys and liver
- Presence of alcohol, drugs, medications, and their possible combined effects on the body when in conjunction with the toxin
- Cumulative effects of previous exposures

5B. Values of Risk Assessment

In determining the effects of toxins on humans, several published standards and values that may be used as resources. The terms a responder may encounter and should understand are defined in the following.

Immediately dangerous to life and health (IDLH). This value, which is published by the National Institute of Occupational Safety and Health (NIOSH), is used primarily in determining the required level of respiratory protective equipment. It is generally taken as the maximum concentration to which a person may be exposed for 30 minutes and still escape in the event of failure of respiratory protective equipment, without suffering any escape-impairing effects. From a selection perspective, atmospheres above the IDLH level must be entered wearing atmosphere-supplying respiratory equipment; atmospheres below IDLH may be entered wearing equipment offering a lower level of protection (see Chapter 3 for further details).

Threshold limit value—ceiling (TLV—C). This is the maximum level that a person should be exposed to even instantaneously. The value is published by the American Conference of Government and Industrial Hygienists (ACGIH). From a practical standpoint, this value has limited application in spill response.

Threshold limit value—short-term exposure limit (TLV—STEL). This is the maximum level that a person may be exposed to for 15 minutes in any one hour, up to four times per day. The value is published by the ACGIH. While this level may seem to have some practical application in spill response, entry into atmospheres at this level should be contemplated only when no other alternatives are available.

Threshold limit value—time-weighed average (TLV—TWA). This is the maximum concentration that a person may be exposed to for 8 hours a day, 7 days a week, without feeling any ill effects. It is the level at which no respiratory protection is normally necessary (exceptions may include operator sensitivity to a substance that dictates a lower exposure).

Permissible exposure limit (PEL). This is the legally published value as enforced by OSHA.

Recommended exposure limit (REL). This is the recommended exposure limit as published by NIOSH.

Several other jurisdictions, notably the provinces in Canada, also set values and limits for exposure. The majority are based on one or more of the above. Table 4.2 summarizes the values defined above and the authorities that set them.

When working in an area where the concentration is below the TLV—TWA, and air-supplying equipment is not being worn, the area must be monitored constantly to ensure that the concentration does not rise. At spill scenes the likelihood of this occurring is high. Many response teams use a safety margin to allow for changing conditions; i.e., if the TLV—TWA for a spilled product is 20 ppm, then the level for unprotected exposure is set at 5 ppm. When circumstances are such that the airborne concentration of contaminants cannot be guaranteed, an atmosphere-supplying respirator must be used.

One other value that may be encountered is LC50 or LD50. These values, *lethal concentration* and *lethal dose,* are the concentrations required to induce a 50 percent kill of laboratory test animals after exposure to a toxin. The value has no direct relation to health effects on humans. Indeed, for some substances, humans have significantly lower tolerance than LC50, while for others the tolerance is higher. The values are used only to compare relative potencies of toxins. As an example, phosgene has an LC50 (rat) of 50 ppm and an LC50 (human) of 3200 ppm.

The LC50 value has sometimes been used as an entry ceiling for air-purifying respirators. This practise is totally without premise and may lead to loss of life. It cannot be overemphasized that the only "safe" value for unprotected exposure is TLV—TWA. Table 4.3 illustrates that there is no direct correlation among TLV, IDLH, and LC50.

TABLE 4.2 Comparison of LC50, IDLH and Threshold Limit Values

Name of Substance	TLV (PPM)	IDLH (PPM)	LC50 (PPM)
Hydrogen Cyanide	5	50	300
Hydrogen Sulphide	10	300	600
Sulphur Dioxide	2	100	1000
Chlorine	1	30	1000
Carbon Monoxide	35	1500	1807
Ammonia	25	500	10000
Carbon Dioxide	5000	50000	10%
Methane	90000	—	?

TABLE 4.3 Standards and Values used in Atmosphere Characterization

Term	Explanation	Authority
I.D.L.H.	Immediately Dangerous to Life and Health. The maximum concentration from which a person could escape within thirty minutes without suffering escape-impairing or irreversible health effects. This value is primarily designed for use in respirator selection.	NIOSH
T.L.V.—C	Threshold Limit Value—Ceiling. The lowest concentration that a person should not be exposed to, even instantaneously.	ACGIH
T.L.V.—S.T.E.L.	Threshold Limit Value—Short Term Exposure Limit. A fifteen minute exposure.	ACGIH
T.L.V.—T.W.A.	Threshold Limit Value—Time Weighted Average. The average concentration for a normal eight hour work day.	ACGIH
P.E.L.	Permissible Exposure Limit. Similar to T.L.V.	OSHA
R.E.L.	Recommended Exposure Limit. Similar to T.L.V.	NIOSH

5C. Toxic Actions

Even though extensive personal protective and air-monitoring equipment is available for use at spill sites, exposures to toxins still occur. Therefore, it is important that response personnel have an understanding of the actions of various toxins on humans. The following physiological categories of toxins are recognized.

a. Irritants

Irritants are substances that have a corrosive action on the mucous membranes. Exposure to such toxins is evidenced by secretion of moisture in the nose and throat and by coughing. Irritants affect the lungs in a similar manner as they effect the mucous membranes of the nose and throat. The lungs have an inadequate mechanism for removing excess fluid, so fluid will accumulate in the lungs. This causes an effect similar to drowning, known as pulmonary edema, a condition that if not treated can be fatal as it prevents adequate oxygen transfer at the cellular level.

As a rule of safety, if a person is exposed to an irritant, assume the "worse case," i.e., that pulmonary edema has been induced, and send the person for medical monitoring. The early stages of pulmonary edema may not be readily apparent to the casual observer. If untreated, possible complications such as pneumonia may occur.

Signs and symptoms of pulmonary edema are:

- Coughing
- Altered mentation
- Restlessness
- Increased respiration
- Increased blood pressure
- Classical white frothy sputum, may be blood tinged
- Cyanosis (blue nailbeds, mucous membranes, e.g., lips)

Administration of oxygen is probably the most effective first aid that can be administered until medical attention can be obtained.

Irritants are divided into three classes:

- Upper respiratory irritants are highly water soluble and are removed in the upper respiratory tract (mouth, nose, and throat). Examples are aldehydes, ammonia, mineral acids, and sulfur oxides.
- Lower respiratory irritants are only slightly water soluble and may be carried throughout the lungs. Examples are cholropicrin, nitrogen oxides, phosgene, and arsenic trichloride.
- Whole respiratory irritants are moderately water soluble and hence are only partly removed in the upper tract. Their effects are severe throughout the tract. Examples are chlorine, toluene, xylene, and dimethylsulfate.

b. Nonchemical Asphyxiants

Nonchemical asphyxiants are substances which interfere with the oxidation process of the body. Closely associated with asphyxiants is the term *anoxia*—a deficiency of oxygen in the body tissues. People will not be aware that their body is not receiving sufficient oxygen unless they are able to recognize the subtle effects of acute anoxia. These progressive effects are:

- Mild euphoria, altered mentation, and confusion
- Increased respiration and heart rate
- Headache
- Fatigue
- Nausea
- Respiratory arrest
- Cardiac arrest
- Death

Substances that cause these results impact by displacing air and hence oxygen. Examples are carbon dioxide, hydrogen, nitrogen, and other inert gases.

c. Chemical Asphyxiants

Chemical asphyxiants are substances that in some way prevent the transfer of oxygen in the body. Examples are:

- Carbon monoxide—has an affinity for hemoglobin that is 200 times stronger than that of oxygen and thereby inhibits the uptake of oxygen
- Anilines—oxygen is not released from the hemoglobin
- Hydrogen sulfide—halts respiration
- Hydrogen cyanide—blocks oxygen uptake by the cells

d. Poisons

Poisons are broken down into two groups: nerve poisons and systemic poisons.

Nerve Poisons
Nerve poisons that are depressants interfere with the transmission of nerve pulses, causing a narcotic effect which can lead to unconsciousness and coma. For hydrocarbon-type substances, the relative strength of narcotic action is:

- Alkynes
- Alkenes
- Ethers
- Propane
- Ketones
- Alchols

Substances that contain chlorine, bromine, iodine, or fluorine are generally more virulent in their effect.

Convulsants act in the spaces between nerve cells, causing continuous stimulation. Exposure is evidenced by uncontrolled motor function. The primary substances in this class are these included in the general category of organophosphorus pesticides.

Systemic Poisons
Systemic poisons cause an effect on specific systems or organs within the body. Their effect is normally a chronic consideration rather than acute.

- Hepatic poisons. The liver is a prime organ for the metabolism of toxins in the bloodstream. In performing this function, the liver is exposed to the toxin and may itself be damaged. Examples of substances that are known to cause liver damage include carbon tetrachloride, acetaminophen, arsenic, and beryllium.
- Nephrotic poisons. The kidneys perform a similar function to the liver,

but there is a mechanism for further removal, in the form of urea secretions. Halogenated hydrocarbons (e.g., methylchloride or chloromethane, chloroform, carbon tetrachloride) accumulate in the kidneys and may cause a carcinogenic response.
- Neurotoxins. The central nervous system (brain, spinal chord, and nerves) is impacted by many dissolved heavy metals, particulary mercury and lead.
- Cardiotoxins. The primary action of poisons on the heart is to sensitize the heart muscle to adrenaline. Once the heart is sensitized, sudden excitement and subsequent release of adrenaline may lead to fibrillation. This action is caused mainly by chlorinated and bromated hydrocarbons.

e. Other Effects
In addition to the preceding, other considerations of exposures involve individual sensitivity to substances, allergenic reactions, and personal sensitization. For example, a person who has suffered repeated low-level exposure to substances such as isocyanates may develop a sensitized reaction that occurs at concentrations well below the TLV—TWA. Similarly, allergic reactions may occur at any concentration.

A final complication to consider is the effect of two or more chemicals in combination that are absorbed into the body. The method in which the body responds to individual chemicals is cataloged and understood. How the body reacts to one chemical in the presence of another is often more difficult to assess. In extreme cases the body may not be able to withstand the effects of two or more chemicals in combination. This effect is known as *synergism*. In a response situation involving mixtures of chemicals, it is wise to assume that a synergistic effect may occur unless there is clear information to the contrary. In these circumstances, the recommended level of chemical protective clothing is Level A.

6. OXYGEN DEFICIENCY/ENRICHMENT

The previous section discussed the effects of oxygen-deficient atmospheres and indications of exposure to them.

There are several circumstances in which an oxygen-deficient atmosphere may arise. The most obvious is the presence of compressed inert gases (i.e., nitrogen). Generally, inert gases are heavier than air, and oxygen-deficient air is denser than normal air. Inert gases that are lighter than air include helium and argon. Whenever an incident involves a confined space, the issue of oxygen deficiency should always be considered.

When wearing atmosphere-supply respiratory protective equipment (SCBA or SAR with egress), the problem is of less significance than when wearing air-purifying units (APRs). Whenever APRs are used, the oxygen concentration should be carefully monitored to ensure a level of at least 19.5 percent, particularly when gases are known to be involved in an incident.

Oxygen-enriched atmospheres may be encountered at incidents involving either compressed oxygen or other materials that release oxygen upon exposure to air, moisture, or other chemicals. These materials are generally included in dangerous goods classes 4 and 5, and wherever an incident occurs that involves such products, an oxygen-enriched atmosphere should be suspected.

Oxygen vapors are heavier than air, and similar to oxygen-deficient atmospheres, investigation should be aimed at lower-lying areas.

The primary risk of oxygen enrichment is enhanced combustion of materials. Not only will the flammable range of vapors be significantly increased, but flammable solids, liquids, and substances liable to spontaneous combustion will all exhibit greater tendencies to burn. When combustion occurs, it will be more rapid and vigorous. Oxygen reacts violently with many commonly used chemicals and hydrocarbon solvents.

From a health perspective, exposure to oxygen at concentrations above 25 percent may cause irritation of the respiratory tract and dizziness.

7. HAZARD INDICATION

There are several systems in use, at national and international levels, that indicate the hazard threat of materials. From an emergency response perspective the most commonly encountered system is the U.N. Dangerous Goods classifications system. Variations, adoptions, and adaptations of this system are promulgated by many bodies and documents. Some examples are:

- International Civil Aviation Organization (ICAO)
- International Air Transport Association (IATA)
- Association of American Railroads (AAR)
- Canadian Transport Commission (CTC)
- Title 49 of the U.S. Code of Federal Regulations (49 CFR)
- Transportation of Dangerous Goods Act (TDGA, Canadian)

These and other documents refer to nine fundamentals classes of dangerous goods/hazardous materials:

1. Explosives
2. Gases

3. Flammable liquids
4. Flammable solids
5. Oxidizers
6. Poisonous and Infections
7. Radioactives
8. Corrosives
9. Miscellaneous

Once a material has been identified by its class, responders can make some initial decisions regarding an appropriate course of action. Further guidance on initial response techniques can be found in publications such as the U.S. Department of Transportation's *Emergency Response Guide Book* and Transport Canada's *Dangerous Goods Guide to Initial Emergency Response*.

The following text discusses the nine classes, their divisions, and applicable markings. It is based on the U.N. classification system.

7A. Class 1: Explosives

The safety marks for Class 1 are black on orange and are shown in Figure 4.11. The general heading for Class 1 is "explosives." Within the class heading there are five divisions:

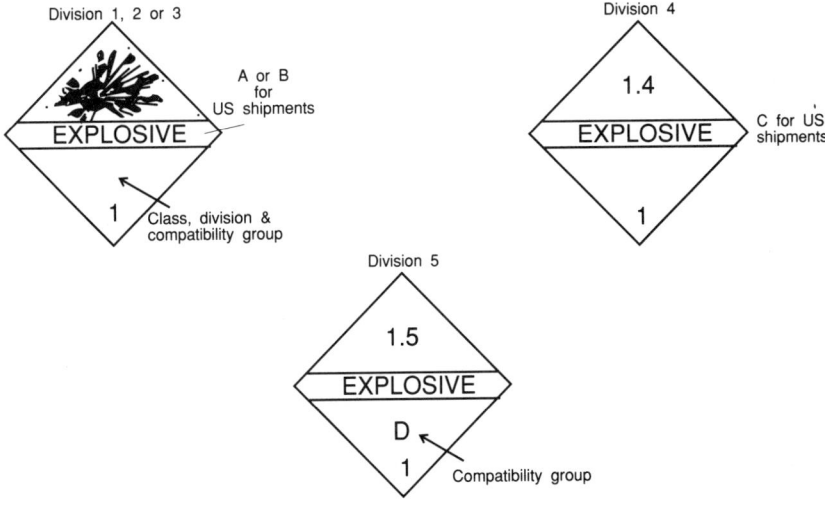

FIGURE 4.11. Class 1 Safety Marks

1.1 Mass explosion hazard—sensitive explosives, e.g., trinitrophenol (U.N. 1344) and trinitrotoluene (U.N. 1356)
1.2 Projection hazard, e.g., cannon charges (U.N. 0414)
1.3 Fire hazard of minor blast of projection hazard
1.4 Explosion effects confined to immediate surroundings
1.5 Mass explosion hazard—very insensitive explosive, e.g., ethylene glycol mononitrate solution (U.N. 0081)

Additionally, 12 compatibility group letters provide guidelines for stowage, segregation, and transportation.

In the United States, Class 1 is divided into divisions A, B, and C. However, the U.N. system of classification will shortly be adopted.

7B. Class 2: Gases

The labels and placards for Class 2 are as shown in Figure 4.12. The general heading for Class 2 substances is "Gases." Products included in this class are those that:

- Have a critical temperature less than 50°C (122°F) or an absolute vapor pressure greater than 294 kPa (43 psi) at 50°C (122°F)
- Exert an absolute pressure in the cylinder, packaging, tube, or tank in which they are contained greater than 275 kPa (40 psi) at 21.1°C (69.9°F)
- Are flammable liquids that have an absolute vapor pressure of more than 275 kPa at 37.8°C (100°F)
- Are gases in the refrigerated liquid form that have a boiling point less than $-84°C$ ($-119.2°F$) at 101.325 kPa absolute pressure
- Are liquid carbon dioxide

Within the class there are three divisions, four in Canada. The divisions are flammable gases, poisonous gases, corrosive gases (Canada), and compressed gases (i.e., those that do not meet the criteria of any of the other divisions).

- Class 2, Division 1: Flammable Gases. These are ignitable at normal atmospheric pressure when in a mixture of 13 percent or less by volume or have a flammability range of at least 12 percent. Examples are butane, propane, and hydrogen.
- Class 2, Division 3: Poisonous Gases. These are gases that have an LC50 value of less than 5000 ml/m^3 at normal atmospheric pressure by reason of toxicity. Examples are methyl bromide and phosgene.
- Class 2, Division 4: Corrosive Gases. These gases have an LC50 value

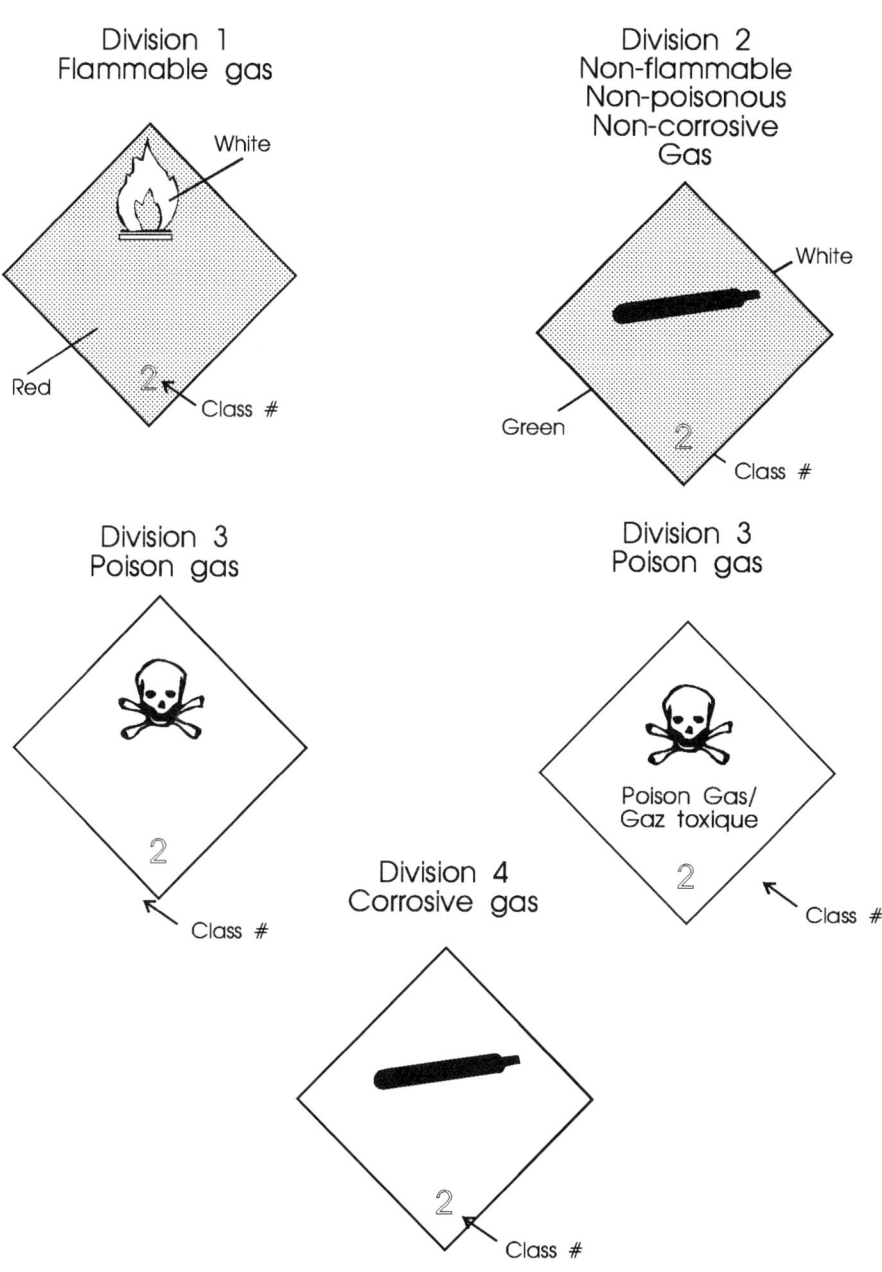

FIGURE 4.12. Class 2 Safety Marks

less than 5000 ml/m³ at normal atmospheric pressure by reason of corrosive effects on respiratory tissue. N.B.: This division is designated only for Canada. Examples are chlorine and ammonia.
- Class 2, Division 2: Compressed Gases. If not included in one of the divisions described above. An example is nitrogen.

7C. Class 3. Flammable Liquids

The labels and placards for Class 3 are as shown in Figure 4.13.

Class 3 dangerous goods are flammable liquids. They are broken down into three divisions and three packing groups. A packing group is defined as "an indication of the inherent level of danger of a product." Packing group I represents the most dangerous form of a class or division and packing group III the least. The division is dependent on flash point and the packing group on boiling point and flash point. All liquids having a flash point of less than 61°C (141.8°F) are flammable liquids, except in the United States, where liquids with a flash point below 37.8°C (100°F) are flammable liquids, and liquids with a flash point between 37.8°C (100°F) and 93.3°C (200°F) are combustible liquids.

The division is determined by flash point as follows:

- Division 1, flash point less than −18°C (−0.4°F)
- Division 2, flash point not less than −18°C (−0.4°F) but less than 23°C (73.4°F)

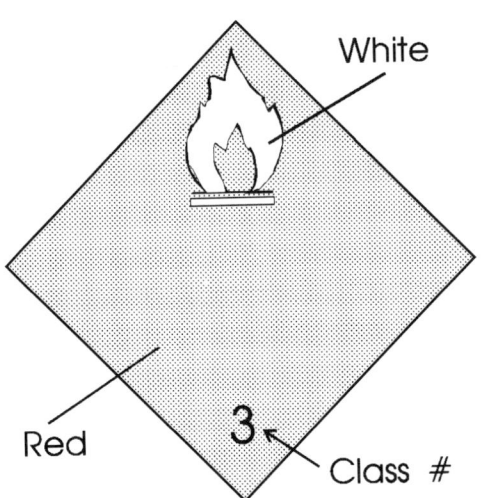

FIGURE 4.13. Class 3 Safety Marks

7. Hazard Indication 97

- Division 3, flash point less than 61°C (air).

The packing group is determined by boiling point as follows:

- Packing Group I, initial boiling point of 35°C (95°F) or less at an absolute pressure of 101.325 kPa
- Packing Group II, initial boiling point greater than 35°C (95°F) at an absolute pressure of 101.325 kPa and a flash point less than 23°C (73.3°F)
- Packing Group III, initial boiling point greater than 35°C (95°F) at an absolute pressure greater than 101.325 kPa

7D. Class 4: Flammable Solids

The labels and placards for Class 4 are as shown in Figure 4.14. There are three divisions in Class 4:

- Division 1: Flammable Solids. These are substances that will readily ignite or undergo self-reaction. As with other divisions in this class, some are so self-reactive that they are assigned an additional, albeit subsidiary, risk of explosive. Examples are silver picrate (U.N. 1347), silicon powder (U.N. 1346), and hay (U.N. 1237).
- Division 2: Substances Liable to Spontaneous Combustion. Substances in this division tend to be susceptible to elevated temperatures and/or ignite upon exposure to air or catalytic gases. For example, tributylaluminum (U.N. 3051) will ignite upon exposure to air or carbon dioxide. Obviously, these substances may prove difficult to extinguish using traditional fire-fighting techniques. Other examples are methyl magnesium bromide (U.N. 1928) and potassium sulfide (U.N. 1382).

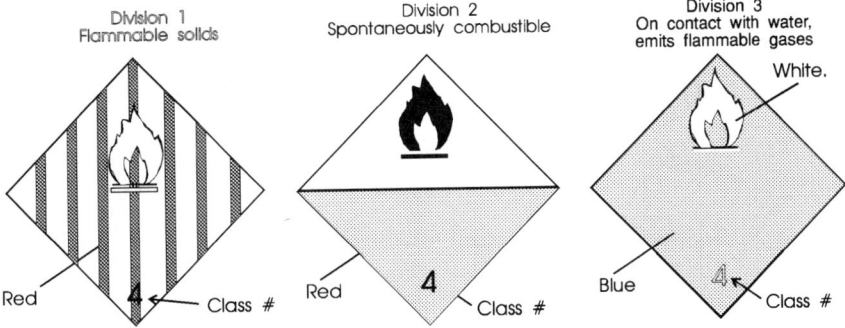

FIGURE 4.14. Class 4 Safety Marks

98   Incident Risks and Safety

- Division 3: Substances That Emit Flammable Gases upon Contact with Water. Many of these substances are so highly volatile that the moisture content of air is sufficient to excite them. Some are transported in a medium designed to specifically exclude contact with water. Examples are cesium metal (U.N. 1407) and ethyldichlorosilane (U.N. 1183).

7E. Class 5. Oxidizers and Organic Peroxides:

The labels and placards for Class 5 are as shown in Figure 4.15. There are two divisions in Class 5:

- Division 1: Oxidizers. These are substances that yield oxygen or other oxidizing substances and thereby contribute to the combustion of other materials. Examples are thallium chlorate (U.N. 2573) and sodium bromate (U.N. 1494).
- Division 2: Organic Peroxides. By definition, these are organic compounds that contain the bivalent O—O structure. The risks attached to these substances are considerable. Most organic peroxides will decompose exothermically in elevated temperatures, and many will explode. Because they are inherently combustible and also yield oxygen, they are extremely difficult to stabilize once decomposition has begun. Indeed, with many substances of this class, evacuation is a good initial action. Many of these substances have low self-accelerating decomposition temperatures (SADTs) and maximum safe transport temperatures (MSTTs). Examples are cumyl peroxyneodecanoate (U.N. 2963) and cyclohexanone peroxide (U.N. 2117).

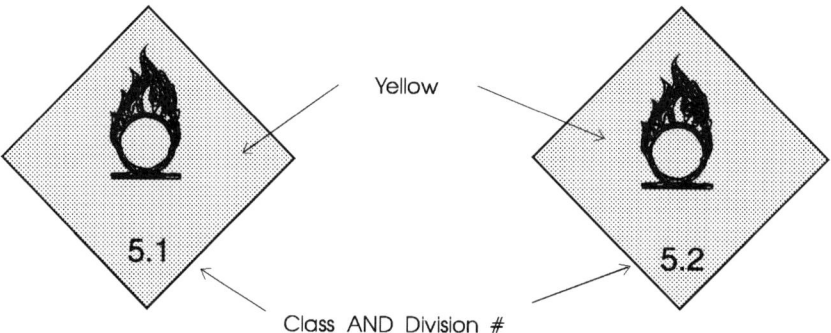

FIGURE 4.15.   Class 5 Safety Marks

7. Hazard Indication   99

7F. Class 6: Poisonous and Infectious Substances

The labels and placards for Class 6 appear in Figure 4.16. Class 6 dangerous goods are broken down into two divisions.

- Division 1: Toxic Substances. These are substances that are defined as poisonous (toxic). Criteria for inclusion pertain to the degree of toxicity that a substance exhibits in laboratory testing. This degree is defined by the LC50 or LD50, where 50 refers to the concentration or dose required to effect a 50 percent kill of test animals over a period of time.

  Within Division 1 of Class 6 there is a further subdivision, packing group. Packing group I represents the most toxic type of product and packing group III the least. In most cases packing group III products must be ingested to cause harmful effects. Examples of this division are arsenic acid (U.N. 1553, packing group I), dinitro cresol (U.N. 1598, packing group II), and arsenical pesticides (U.N. 2994, packing group III).

- Division 2: Infectious Substances. Infectious substances are essentially any organisms that are infectious or reasonably believed to be infectious

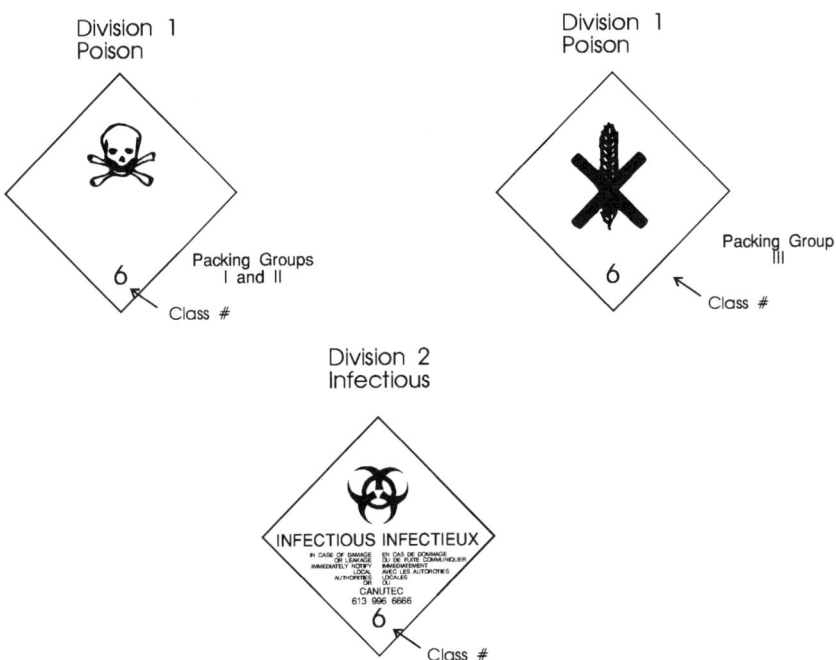

FIGURE 4.16.   Class 6 Safety Marks

100   Incident Risks and Safety

to humans or animals. The principal consideration for these substances is that traditional personal protective equipment, operating procedures, and decontamination protocols may not be sufficient to safely resolve problems involving these substances. Before approaching any product included in this division, responders should contact CANUTEC or CHEMTREC for advice.

7G. Class 7: Radioactives

There are three labels and one placard for Class 7 dangerous goods. The labels and placards are as shown in Figure 4.17. Before dealing with incidents involving radioactive materials, responders must have further

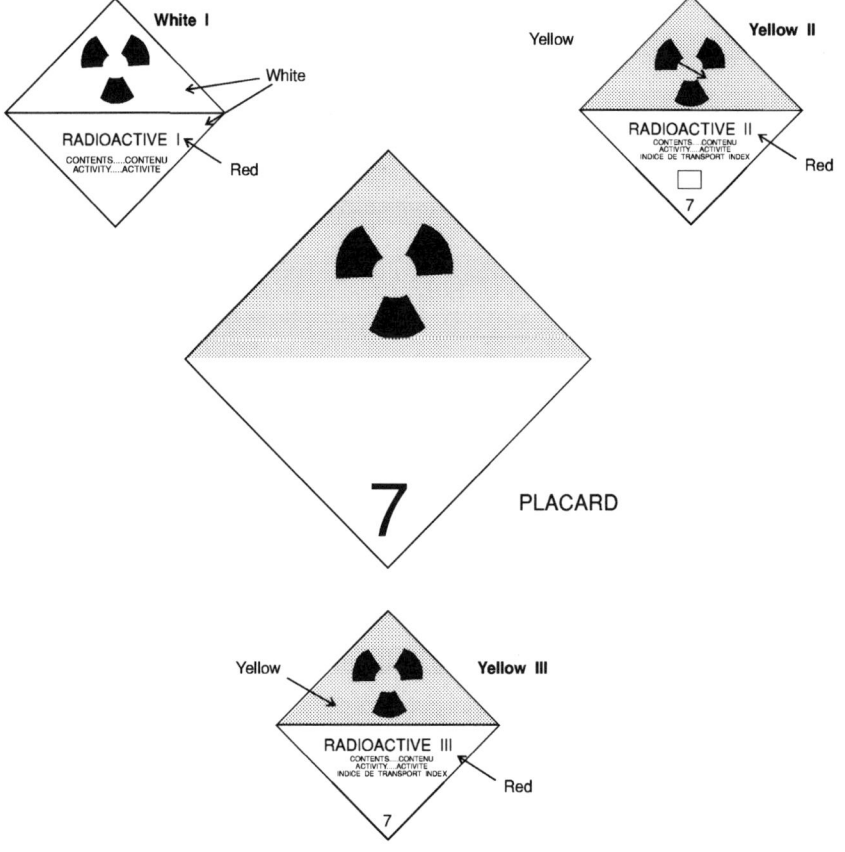

FIGURE 4.17.   Class 7 Safety Marks

FIGURE 4.18. Class 8 Safety Marks

training relating specifically to this field. This training is beyond the scope of this book. Fortunately, most shipments of radioactive materials are accompanied by a trained response team.

7H. Class 8: Corrosives

There is one label and one placard for Class 8 dangerous goods, as shown in Figure 4.18. Class 8 dangerous goods are dangerous goods that:

- Have been known to produce visible damage to human skin tissue
- Cause visible damage on the skin tissue of an albino rabbit within 4 hours of application
- Corrode SAE 1020 steel within the parameters set out in the regulations
- Are wastes that have a pH less than 2.0 or greater than 12.5

There are no divisions in Class 8, but there are three packing groups, all of which depend on the time required to produce the conditions outlined for Classes 2 and 3. Examples are sulfuric acid (U.N. 1831, packing group I), acrylic acid (U.N. 2218, packing group II), and aluminum chloride solution (U.N. 2581, packing group III).

7I. Class 9: Miscellaneous

The label and placard for Class 9 is shown in Figure 4.19. Class 9 covers miscellaneous products and substances, i.e., those that do not meet the criteria defined for inclusion in other classes. There are three divisions in Class 9.

- Division 1: Miscellaneous Dangerous Goods. Examples of this division are polychlorinated biphenyls (PCBs), which are not poisonous, corrosive, explosive, or flammable within the criteria of this classification

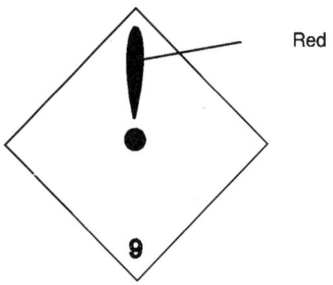

FIGURE 4.19. Class 9 Safety Marks

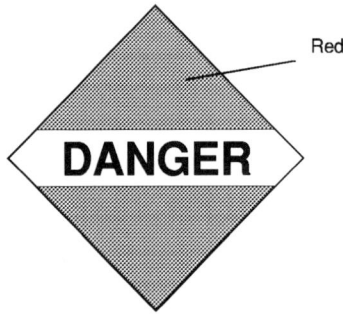

FIGURE 4.20. Danger Placard

system. However, they still present a significant environmental risk if not handled correctly.
- Division 2: Hazardous to the Environment. The primary threat posed by these substances is to flora, fauna, and other environmental considerations. An example is formaldehyde solution (U.N. 2209). Additionally, many substances are assigned this division as a subsidiary risk.
- Division 3: Dangerous Wastes. These are cataloged wastes from specific industrial processes, for example, waste type 6 (N.A. 9306).

There is one other safety mark that is used in this system and is significant to responders. That is the danger placard, as shown in Figure 4.20. Its presence on a vehicle indicates that the unit contains a mixed load of hazardous materials in varying quantities—for example, drums of gasoline, sulfuric acid, and paint. When this safety mark is encountered, the first objective, after searching for life, should be to secure the shipping document and glean from it the exact nature of the goods carried.

This section has reviewed one system of hazard indication and shown the main groupings associated with it. Responders should be aware that substances may have more than one risk associated with them. For example, phosphorus (U.N. 2447) is spontaneously combustible (Class 4, Division 2), poisonous (Class 6, Division 1), and harmful to the environment

(Class 9, Division 2). It is therefore important that safety marks be used only for initial-response decision making; all products involved must be clearly identified as soon as possible.

8. OTHER INCIDENT RISKS

In addition to the risks already addressed, several others should be considered at an incident site. These include:

- Fire
- Terrain
- Loose equipment
- Slips, trips, and falls

8A. Fire

The issue of fires involving hazardous materials is covered in other volumes. Although this book is designed for spill response rather than fire fighting, some of the complications involved are worth highlighting.

One of the most difficult issues is personal protection. Most chemical protective clothing (CPC) has little, if any, fire-retarding qualities. Furthermore, in the presence of excessive heat they lose many of their chemical-resistant qualities. Protective clothing is available that meets both protective requirements, but it is costly, bulky, and needs significant operator training in its use.

A further complication of this issue is that in a fire, especially in the early stages, it is very difficult to determine the substances involved; couple this with the combined effects of chemicals exposed to high temperatures, and it can be appreciated that CPC selection is difficult. Generally, personal protection is aimed at protection from the fire itself and protection of the respiratory system by use of SCBA.

If those involved in fighting fires experience skin discomfort that may be due to chemical exposure, the most prudent approach is to withdraw and allow the fire to burn, rather than risk exposure of personnel.

During fire-fighting operations, large volumes of runoff water are created. As far as is practicable, this water should be collected, analyzed, and treated. In any event, after a fire involving hazardous materials, a full environmental survey for material impact on soils, groundwater, structures, etc., will be necessary.

Many hazardous materials exhibit properties that make fire fighting difficult and require specific techniques and extinguishing materials. As discussed previously, hazardous materials may emit oxygen, explode, react with water, emit poisonous gases, etc., and all these risks are amplified in fire fighting situations.

8B. Terrain

Often it is necessary to set up a command post at some distance from the actual spill site. When the terrain is flat, observation of an approach team is reasonably straightforward. However, with undulating terrain the approach team may not be visible at all times. Where materials are involved that may lead to the evolution of dense gases, such gases may tend to accumulate in the lower-lying areas of the terrain. Additionally, if a team member slips or falls in a low-lying area, it may not be immediately apparent to the command team. For these reasons, whenever an approach is made over uneven terrain, constant radio communication is essential. Should this not be possible to facilitate, the only alternative is to use "spotters" within the exclusion and contamination reduction zones. Obviously, such personnel will need to be equipped with the same personal protective equipment as the approach team.

Other terrain-related problems relate to slips and trips on potholes, curbstones, etc. It cannot be overemphasized that the incident command team must at all times be able to maintain contact with an approach party. Ideally, this contact will be visual.

8C. Loose Equipment

At transportation incidents, loosened packaging and unstable transport units pose a significant threat. This must be one of the first priorities for assessment; it is pointless to determine that no threat exists from hazardous vapors and then have a person rendered unconscious by a falling cylinder.

In plant situations the usual exposures, such as falling tools, etc., apply equally to incident sites as workplaces; indeed, many in-plant incidents occur during maintenance operations, when there are more tools and heavy, loose materials present than normal.

8D. Slips, Trips, and Falls

Personnel dressed out in protective clothing and equipment often have hindered vision and a raised center of gravity; the "bulk" of their equipment means that they occupy more space than usual. Therefore, they are more susceptible to the common slips, trips, and falls to which many of us are prone in the workplace. The problem is further complicated by the protective equipment in that a person will fall harder and find it more difficult to stand after falling. As in the previous section, this points to the need for constant vigilance on behalf of the command team and extreme care on behalf of entry-team members.

9. INCIDENT ASSESSMENT, APPROACH, AND MONITORING

No two incidents are alike, so it is impossible to detail a response for every perceived scenario. However, there are similarities in incidents from the perspective of strategies in command, control, management, and approach.

9A. Initial Assessment

In a plant/warehouse situation at a facility known to the responders, the initial assessment stage can often be readily addressed. Recent legislative moves in the United States (federally, under SARA) and Canada (municipally, by fire departments) have led to large amounts of information being available to responders—too much, some argue. On approaching an industrial setting it is often known what materials may be involved. With unknown industrial settings and transportation incidents, many of the factors discussed herein will be applicable.

The objectives of initial assessment are primarily to establish the risks to life, property, and environment, and to establish initial control zones and an initial location for the command post. The following sequence will aid in meeting these objectives.

a. Indicators

In transportation incidents, look for labels, placards, commodity names, and any other form of labeling/marking. If possible, contact the shipper or transport unit operator and try to glean more information regarding the possible products involved.

Other visual signs that may indicate the depth of the problem are vapor clouds, product staining, the condition of wildlife, etc. Binoculars are an excellent resource at this stage of an incident.

While odor detection is not the ideal method of incident assessment, odors may give an indication of some of the products involved. When odors are encountered, immediate response/reaction may be necessary. For example, if, while surveying a scene, a strong ammonia odor is detected, it would be foolhardy to ignore the odor until the presence of ammonia can be qualified by instrumental observations; it would be equally foolhardy to approach an incident using odor as the only method of detection.

b. Factors

Once some indication has been gleaned regarding the nature of products involved, several other factors need to be considered very early in the incident.

Wind. Is the wind direction or forecast wind direction likely to cause vapor spread in a direction that could possibly harm human or animal life?

Temperature. Will the products involved tend to destabilize or emit flammable/toxic vapors as temperatures vary? If so, what early precaution, such as minimal "safe" distances, are appropriate? In conjunction with wind, temperature will be a major factor in the propagation and spread of vapors.

Precipitation. Any kind of precipitation will serve to hinder response operations. Such hindrance may range from mere inconvenience through reduced visibility, to spread of spilled materials, to destabilization of water-reactive materials. In some cases precipitation may be of an advantage; e.g., if ammonia or chlorine are involved and vapors are present, rain will tend to "knock down" the vapor cloud. Snow and freezing rain tend to cover not only spilled product but also incident hazards, such as loose objects, potholes, etc. In extreme cases, whole transport units have been missed because they have sunk and been covered in snow.

Traffic control. Often, particulary with highway incidents, an incident itself will cause traffic flow to cease. In any event, traffic control should be deferred to the police as soon as they are in attendance.

Terrain and obstruction. As mentioned before, undulating terrain may hide dangers, make it difficult for entry teams to be constantly monitored, and make approach to a site more difficult. Whenever possible, positions such as a command post should be established on the highest suitable ground.

Equipment review. While a review of equipment forms an essential part of the planning stage, it is essential that a final review be made at the incident scene. Particular attention should be paid to the suitability of equipment to dealing with the incident. As more is understood about the incident, it may prove necessary to call on items from sources other than internal ones. The contingency plan should point out where these are available. As others interface at an incident scene, new sources and resources may become apparent. In any event, never approach an incident until all appropriate equipment is available.

Other sources of aid. There are many agencies, both government and industrial, that may be able to render assistance at an incident site. These include, but are not limited to, the organizations shown in Table 4.4.

Possible risk. The final deliberation in incident assessment is that of incident risk, i.e., what threats exist to life, property, and the environment. Subsequent actions and priorities in dealing with the incident will depend on the risks identified. A general tenet is that risks are addressed in the order of life, property, and the environment. There may be several issues within each category, in which case further prioritization may be required.

As an example of the last item above, consider a release of a poisonous gas at an industrial facility. Two people have been injured during the release, the facility population is threatened by the escaping vapors, and there is a large subdivision a short distance downwind. Efforts are being made to control the vapors by using a water fog from an upwind position, the runoff from which is reaching a nearby river. A few dead birds have been found. In assessing and categorizing the risk, the decision-making process would be as shown in Table 4.5.

As can be seen, the process of prioritization can become extremely complex and subjective. The first priority must be the protection of the response team; other life-threatening situations cannot be addressed if the response team is nonfunctional. Following this consideration, the subsequent initial actions may be:

- Recover and evacuate the two victims. They are already injured and will probably suffer the highest exposure at the incident.
- Evacuate all nonessential personnel from the facility to an upwind position.
- Evacuate residents from the downwind subdivision. This will probably occur simultaneously with the evacuation of facility personnel, provided

TABLE 4.4   Risk Categorization

Life	Property	Environment
Response teams	Equipment damage	Runoff water
Victims	Down time	Aquatic exposure
Plant personnel	Leaking tanks	Airborne contaminants
Local population		

108   Incident Risks and Safety

TABLE 4.5   Summary of Sources of Aid

United States	Canada	Both
Federal Emergency Management Agency (FEMA)	Environment Canada	Fire departments
	Provincial Environment Ministries	Police departments
U.S. Environmental Protection Agency (USEPA)	Transport Canada	Paramedics
	Transport Dangerous Goods	
State Environmental Protection Agencies	CANUTEC	
	TEAP	
CHEMTREC	CHLOREP and other product-specific mutual aid groups	
CHEMNET		
CHLOREP and other product specific mutual aid groups		

sufficient resources are available. Personnel sent to the downwind areas may need personal protective equipment.
- Contain runoff water from the vapor-suppression activity.

In this example, several "life" issues have to be categorized and dealt with in a logical sequence. Environmental issues that involve the lives of animals have priority above those that involve contamination of vegetation or property.

9B. Initial Actions

Based on the findings on the initial assessment, the following actions can be initiated. It cannot be overemphasized that extreme caution should be exercised at this stage. Even for a small spill in a plant setting, it is important that the information gleaned from the initial assessment be utilized in deciding on the initial actions. Furthermore, by taking the time to implement these actions, the best possible protection will be afforded. The following steps should be observed prior to actually approaching an incident.

a. Personal Protection

As discussed in Chapter 3, it is always best to err on the side of caution when selecting chemical protective clothing and respiratory-protective equipment. Personal protection is necessary not only for the entry and backup parties but also for those assigned other duties where they may be exposed to harm, i.e., air monitoring, decontamination, etc.

The levels of protection assigned will be based on initial observations and the risk determined from them. Whenever the products involved are

9. Incident Assessment, Approach, and Monitoring

not clearly identified, or the products involved dictate so, Level A is the preferred level of protection; in any event, never approach an incident where unknown products are involved in less than Level B protection.

While caution is of the utmost importance, it must be tempered with reasonableness. For instance, it would be overcautious to approach a rolled-over truck clearly marked with U.N. 1202 placards in Level A (U.N. 1202 being the number assigned to high-flash-point flammable liquids such as diesel oil).

b. Air Monitoring

Once the required level of protection is defined, the initial assessment of vapor concentrations at a site can be made. The exact technique used will depend on factors such as wind direction, topography, etc. Generally, testing is carried out beginning from an upwind position, or in no-wind conditions, from a point deemed sufficiently distant from the incident site.

Initial air monitoring is carried out in the highest level of protection required at the incident scene. Figure 4.21 illustrates a suggested initial monitoring program at an incident site involving benzene. Chapter 6 discusses further the techniques and instruments used in air monitoring.

It should be noted that the primary objective of air monitoring is to help in deciding which areas require the various types of protective equipment. In the initial stages, all entries will be in the highest level appropriate. Once the distribution of airborne contaminants is understood, appropriate adjustments may be made; i.e., if the vapor concentration is clearly below IDLH at all points and the incident dynamics are clearly understood, it may be appropriate to down-stage from SCBA to APRs.

c. Zones

Once initial air monitoring has been effected, the next stage is to establish a more permanent command post and more clearly defined control zones. The areas surrounding an incident are divided into three control zones, as shown in Figure 4.22.

Exclusion zone
The *exclusion zone* should be entered only while wearing the highest level of protection designated for the incident. This is the area where exposure to harmful vapors and liquids may take place, hence equipment and personnel may become contaminated. The exclusion zone is bounded by the exclusion line or hot line. The exact position of the hotline depends on several factors, such as oxygen concentration and the presence of hazardous vapors, topography, distance from the spill site, wind speed and direction. Once established, the hot line should be clearly marked using, for example,

110   Incident Risks and Safety

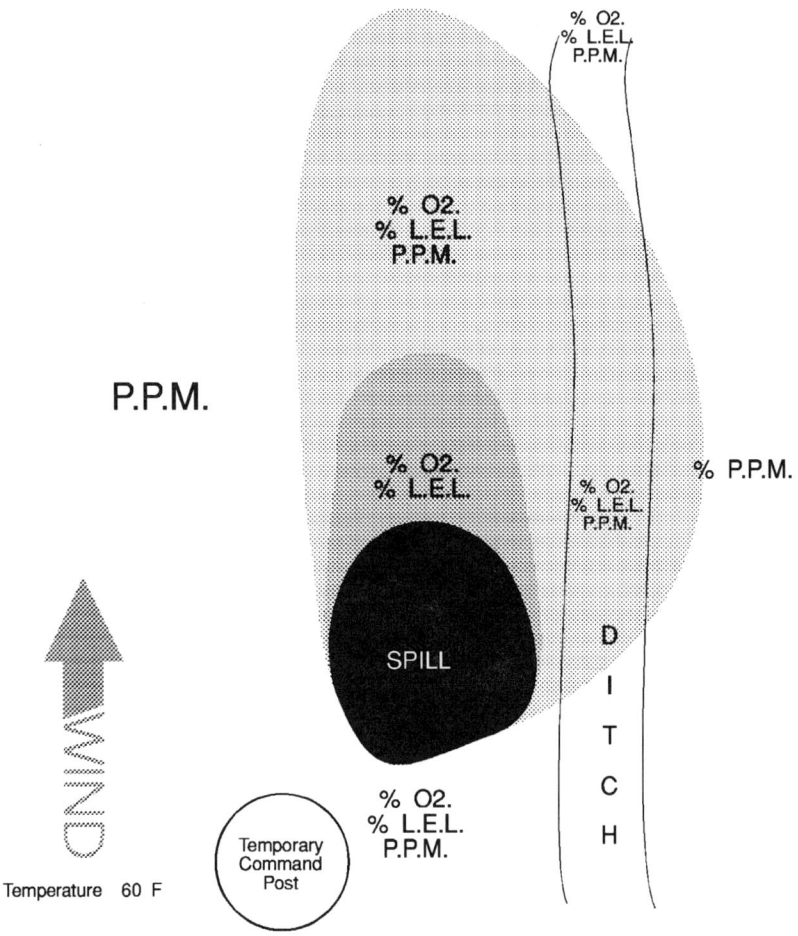

FIGURE 4.21.   Initial Air Monitoring

yellow caution tape or marking dye; where this is not feasible, it should certainly be clearly defined in the minds of all team members, perhaps by the use of topographical features.

Contamination Reduction Zone

The *contamination reduction zone* is designed to act as a buffer zone between the exclusion and support zones. Additionally, along with the contamination reduction corridor, it is the area where decontamination takes place and equipment is resupplied.

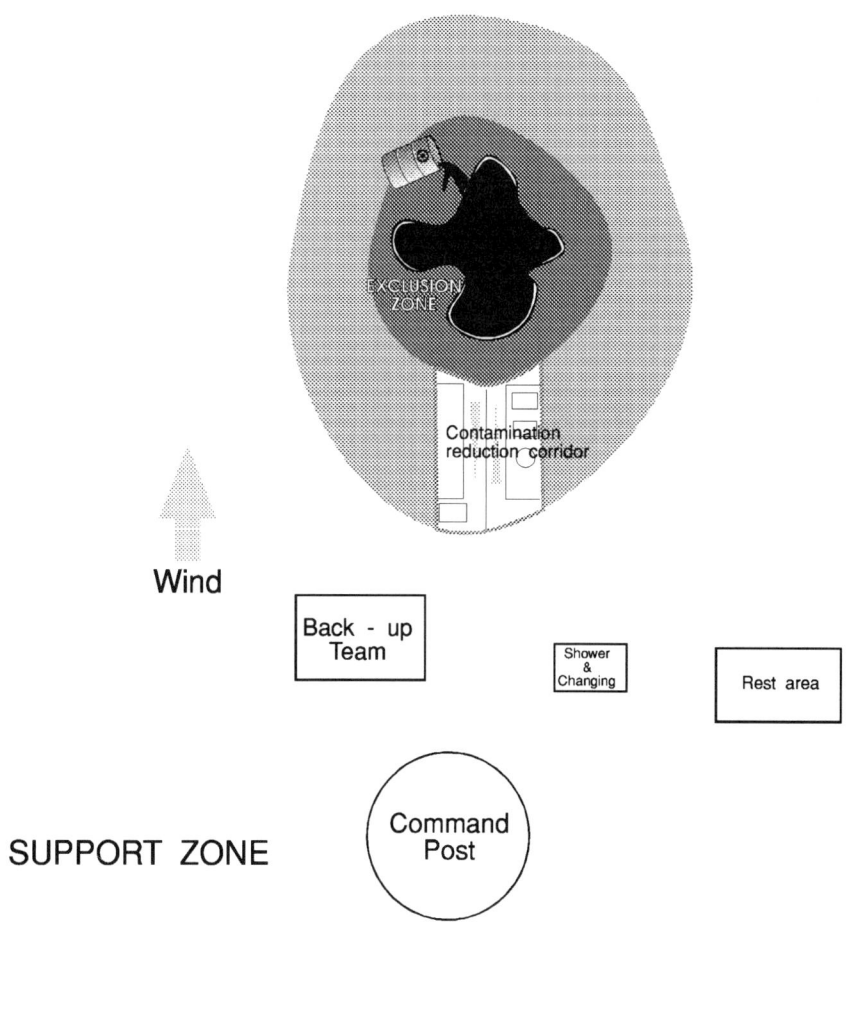

FIGURE 4.22. Spill Site Setup

Personal protective equipment worn in the contamination reduction zone may be of a lower level than that utilized in the exclusion zone, particularly when dealing with gases. Whenever there is risk of exposure from contaminants on an entry party's personal protective equipment, personnel in the contamination reduction zone must wear the same level of protection. The contamination reduction zone is bounded by the support line.

Support Zone

The *support zone* is the area where no contaminants are present in concentrations above the TLV—TWA, and it must be monitored to ensure that this remains so. No protective clothing or equipment is necessary in this area. It is recommended that Level D be adopted as a minimum standard in the support zone. Similarly, no contaminated equipment from the incident site must be introduced into this area. Other activities and facilities in the support zone include the command post and communications center, rest facilities, equipment maintenance, air cylinder refilling, and all other such functions that do not require the use of personal protective equipment. Figure 4.22 shows the interrelation of these zones at an incident site and suggested areas for equipment and facilities location.

9C. First Approach

The objective of the first entry into the exclusion zone is to search for victims, assess the risk of explosivity, and survey the incident scene in detail.

Prior to entry, the entry, backup, and command teams should have discussed and clearly understood the procedures for communication, decontamination, and emergency exit from the spill site.

The entry team approaches the exclusion zone by way of the contamination reduction corridor. For every member of the initial entry team (at least two), there should be a backup person in identical protective equipment standing by in the support zone.

During approach, the entry team should take confirmation readings of airborne contaminants and oxygen concentration. Where possible, a constant monitoring device that assesses oxygen concentration and the presence of combustibles should be used.

Once at the spill/incident site, a thorough search should be made for injured personnel. It is important to search not only the immediate vicinity of the spill but also surrounding areas within the exclusion zone, particularly downwind. Often people who have been involved in an incident will have been able to move some distance away from the site before being overcome by vapors, tripping, or falling. If incapacitated personnel are found, then entry-team members should remove such personnel by way of the contamination reduction corridor to the support zone, where they can be treated.

Other actions to be taken during this first approach include confirmation of any initial observations, a search for any appropriate documentation (i.e., the shipping document), and observation of any hazardous liquid paths and their interception with surface waters. Additionally, some critical, easily accomplished tasks may be undertaken; these should be limited to securing loose objects, rolling leaking drums until the leak is in the vapor area, closing valves, etc. Beyond this, no further action should be taken during the first entry. Once all possible information is gleaned, the entry team should leave the exclusion zone and go to the contamination reduction zone.

Between the first and second entry, the entry team must brief the command team and decisions must be made on the achievable objectives for the second entry. In many cases it will suffice for the briefs to take place with the entry team standing in the contamination reduction corridor, rather than going through full decontamination procedures. If the team is using SCBA, a cylinder change should be affected at this point so that the second entry is started with full cylinders.

9D. Second Approach

The primary objective of the second approach (assuming the objectives of the first approach have been reached) are to further stabilize the incident and continue air monitoring. Hopefully, at the end of this approach all leaking containers will have been identified and as far as possible stabilized.

Until all materials have been fully characterized, it will not be possible to develop a definitive air monitoring protocol. Once all materials have been fully identified and a full atmospheric characterization is achieved, it may be possible to downstage the assigned levels of protection.

9E. Subsequent Approaches

During subsequent approaches the entry teams will move progressively from stabilization to neutralization and cleanup. A primary objective is to stabilize the situation to a point where lower levels of personal protective equipment may be used. This allows for a more expedient progression of incident resolution.

Other tasks to be undertaken at this stage include:

- Cataloging all materials
- Estimation of quantities involved
- Securing containments
- Preparing media presentations
- Arranging extra equipment and personnel
- Setting up a medical monitoring program

As an incident progresses into these phases, extra emphasis must be placed on issues such as monitoring and safety; as personnel become more accustomed to an incident, there is a tendency to overlook these issues and for accidents to occur.

9F. Removal and Restoration

Once the incident is clearly under control and all sources have been identified, the next stage is to deal with removal of hazardous materials and contaminated soil or water. At this stage the response team will probably look to turn over operations to an environmental contractor.

While some previous efforts will have been made to reduce the level of protection required at the incident, a constant assessment of the level of protection in use must be made. When recovery operations start, a whole new set of circumstances may arise due to disturbance of materials, soils, and water; pumping, for instance, can greatly increase the quantity of vapors being developed by a volatile liquid.

It is not within the purpose of this book to discuss techniques of site remediation. However, issues to be resolved at this stage include:

- Estimation of quantities to be recovered and disposed of
- Spill paths and environmental impact
- Soils testing protocols
- Cleanup criteria (How clean is clean?)
- Review of health and safety protocols
- Treatment methods on site and off site
- Disposal
- Permitting, notification, and reporting procedures

5
Spill Control

This chapter deals with the techniques of spill control. The methods and technology available change constantly; new recovery equipment is announced on an ever-increasing basis. While it is beyond the scope of this book to deal with every type of equipment, the most common are addressed.

1. CONTAINMENT

1A. Spill to Land

Land-based spills can cause some or all of the following problems:

- Soil contamination
- Vegetation contamination
- Groundwater contamination
- Runoff to surface waters
- Sewer invasion
- Vapor migration

The exact path that a product spilled to land will take depends on many factors. These are determined by product characteristics (density, viscosity, pour point, vapor pressure), soil permeability, groundwater level from the surface (known as the water table), and sewer proximity. These factors are illustrated in Figure 5.1. The diagram assumes a "worst-case" scenario, i.e., a volatile, low-viscosity product, such as gasoline, spilled in an area of permeable soil with a high water table. Obviously, the longer a situation like this is allowed to exist, the more dramatic will be the impact on the environment.

116   Spill Control

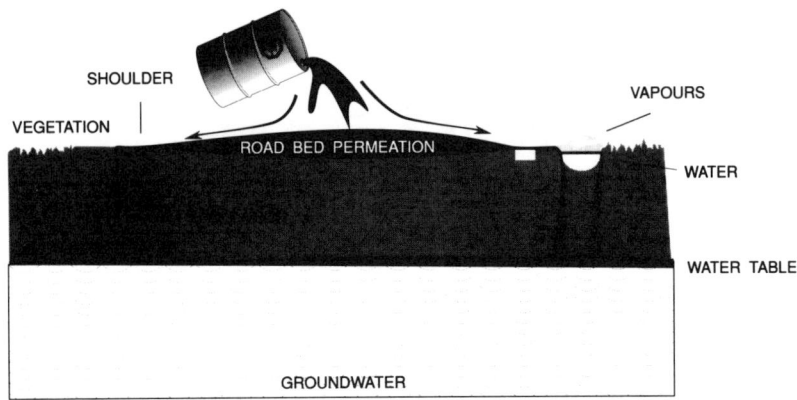

FIGURE 5.1.   Product Spill Paths

No two situations will be the same. Table 5.1 illustrates the effects of temperature, wind, rain, snow, and soil permeability on product movement.

Figure 5.2 illustrates general procedures to be followed in the event of a spill. Specific techniques are discussed below.

Containment is achieved by using one or more of the following methods.

a. Dyking

Dykes are easily constructed using either commercially available units or, more often, surrounding soil and other similar materials. Depending on spill size, dykes are constructed using equipment ranging from shovels to commercial backhoes. When flammable products are to be dyked, great care must be taken to avoid ignition from the electrical components and moving parts of the unit. This often prohibits the use of larger, mechanical units. Dykes should be constructed a safe distance away from the leading edge of a flammable product.

There are two common mistakes made when constructing or laying dykes. One is to attempt to contain too large an amount of product in a given area; the other is to deploy the dyke too close to the leading edge of the spill. The former leads response personnel to build dykes that are too big and unable to withstand the pressures exerted on them by the liquids they contain; the latter causes breaching of the dyke due to incomplete construction or deployment. As a rule of thumb, a dyke should be twice the height of the liquid it is required to contain, with a base that is twice the height of the dyke (i.e., four times the height of the liquid).

Initial efforts at construction of dykes will primarily involve the placement of soil or sand. Response personnel should aim to refine the dyke

TABLE 5.1 Effects of Variables on Product Movement

	Gasoline	Heavy fuel oil	Liq sodium hydroxide	
T E M P	No noticeable effect on mobility. Directly effects vapor generation.	Low temperatures reduce product mobility and rate of percolation.	At low temperatures product will tend to crystallize and stabilize.	T E M P
W I N D	Winds increase rate of vapor spread, also dilution rates.	No noticeable effect except spread of odours.		W I N D
S O I L	Most soils are permeable to gasoline. Only materials such as clay will significantly reduce percolation.	Many soils are impermeable to product. Percolation is much reduced. Soil temperature decreases with depth at upper levels thus stabilizing product movement.	Most soils are permeable to product.	S O I L
R A I N	Product will float on surface accumulations and migrate much faster. Sewer travel is significantly increased.	Tends to cool and stabilize product. Reduces mobility. Some factions water soluble and may be liberated by rain.	May cause some dilution, however, main risk is that of product spread.	R A I N
S N O W	Product still flows freely. Vapours may migrate significant distances under snow.	Rapid cooling of product and hence, stabilization.	Similar effects to rain. In extreme conditions could reduce tendency to crystallize.	S N O W

construction, as circumstances permit. Typically this involves increasing the amount of material in a dyke, adding an impermeable layer (i.e., plastic sheeting), and constructing secondary barriers. Figure 5.3 illustrates these successive stages.

No dyke will ever totally prevent product movement, but significant restrictions and temporary containment can be achieved. Depending on wind conditions and product volatility, dykes may also help to restrict vapor movement.

Once the product is contained, immediate procedures for recovery must be implemented, especially with a low-viscosity product in an area of high soil permeability.

b. Trenching

The construction of interceptor trenches, coupled with the use of existing trenches, is another method of product containment. The exact method of

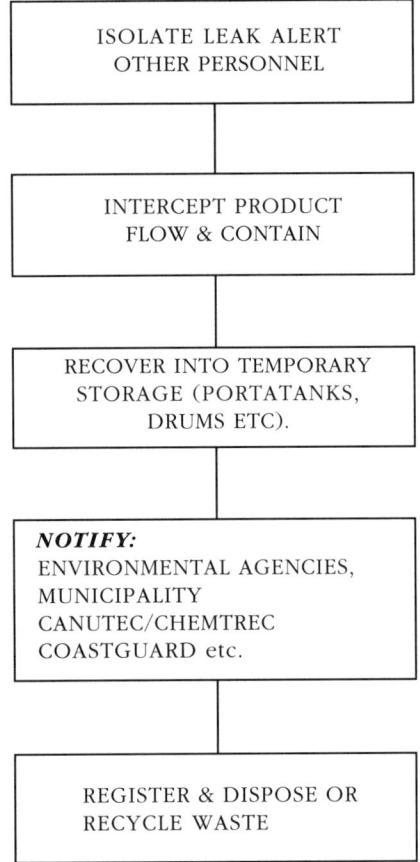

FIGURE 5.2. General Procedures for Spills

construction and maintenance will, again, depend on issues such as soil porosity, product solubility, etc. For example, the most effective method of preventing diesel oil permeating a trench bottom is to allow a certain amount of water to enter the trench. Figure 5.4 shows a typical trench construction for containment of nonsoluble product with a density less than that of water.

If water is not available, then an alternative is to totally line the trench as in Figure 5.5.

Interceptor trenches and dykes may still be useful for nonsoluble products and those with a relative density greater than water, but effectiveness will be significantly reduced.

1. Containment   119

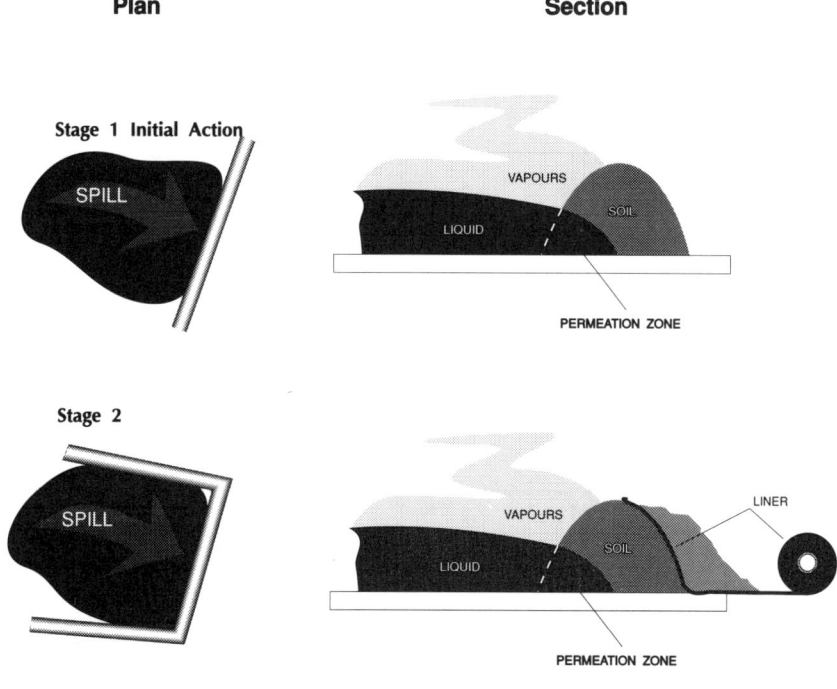

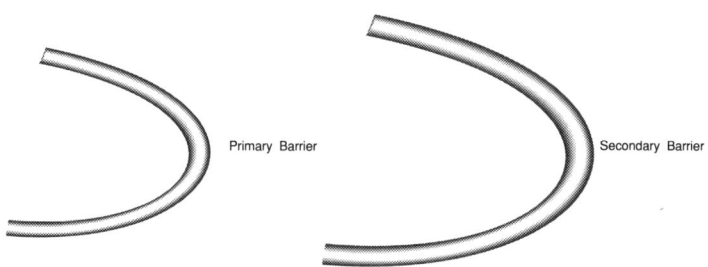

FIGURE 5.3.   Dyking Progression

Once a trench is constructed, be careful to monitor the liquid level. Should the ingress of water, perhaps from precipitation, cause a rise in level, steps must be taken to remove it.

c. Ditches and Small Streams

In many situations spilled product has a tendency to collect in preconstructed ditches. In such circumstances the primary aim is to control the

120   Spill Control

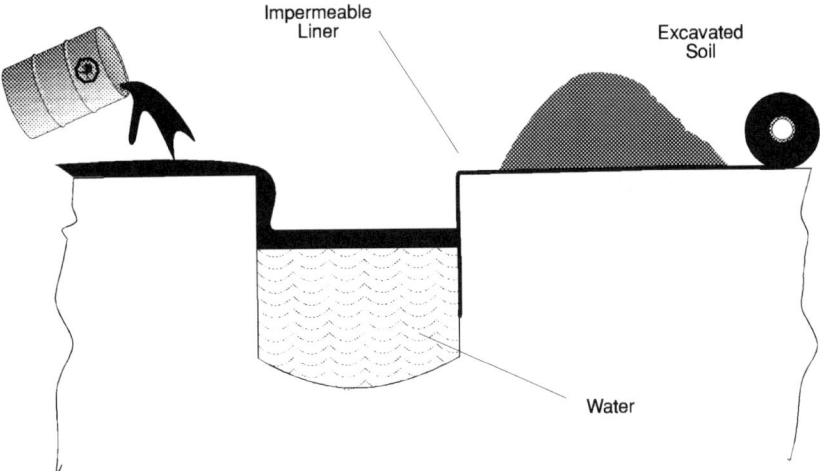

FIGURE 5.4.   Interceptor Trench Construction

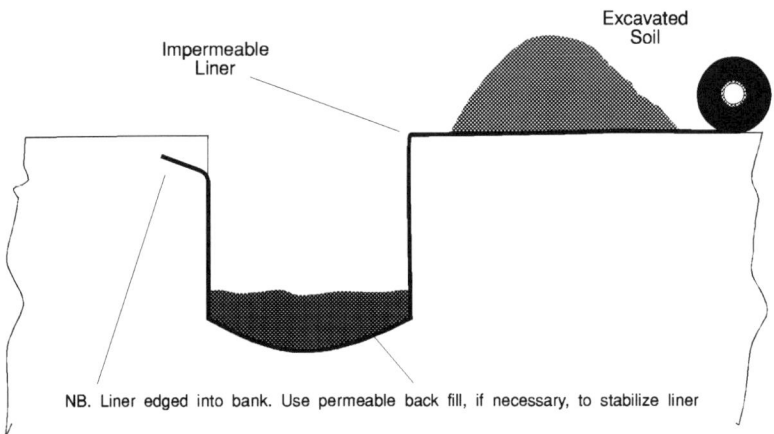

FIGURE 5.5.   Fully Lined Trench

movement of product along the ditch, without hindering the movement of water. This is accomplished by the construction of dams or weir-type arrangements at strategic points. Figure 5.6 shows a typical ditch layout, in this case along a roadside.

Control of product flow may be achieved by use of one or more of the following techniques.

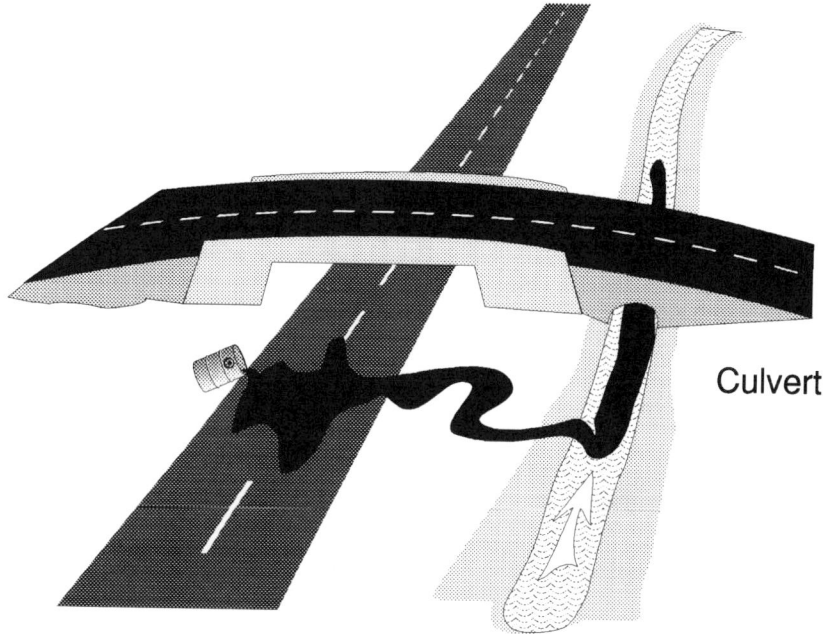

FIGURE 5.6. Typical Ditch Layout

d. Weir

A weir is simply a barrier that impinges the water surface to a depth that is sufficient to prevent the passage of oil. Figure 5.7 shows the construction of a weir using 12-in. boards; other materials, such as logs, may be just as effective. If possible, weirs should be angled across the direction of flow. This will cause product movement to one side of a ditch or stream and reduce the tendency of eddy currents to develop at the weir, which tend to entrain product under the weir.

Most situations require the use of more than one weir to control the product movement. Be careful to ensure that a good seal is achieved at points A and B. Failure to do so will result in breaching of the weir. Sealing can be achieved by using mud, straw, sorbent pads, sheeting, etc.. If possible, plastic sheeting should be used to line the ditch sides/river bank to help reduce contamination.

e. Dams

In narrow ditches and streams, dams are an effective method of controlling product movement. Initially, rocks and/or soil are deposited in the stream

122   Spill Control

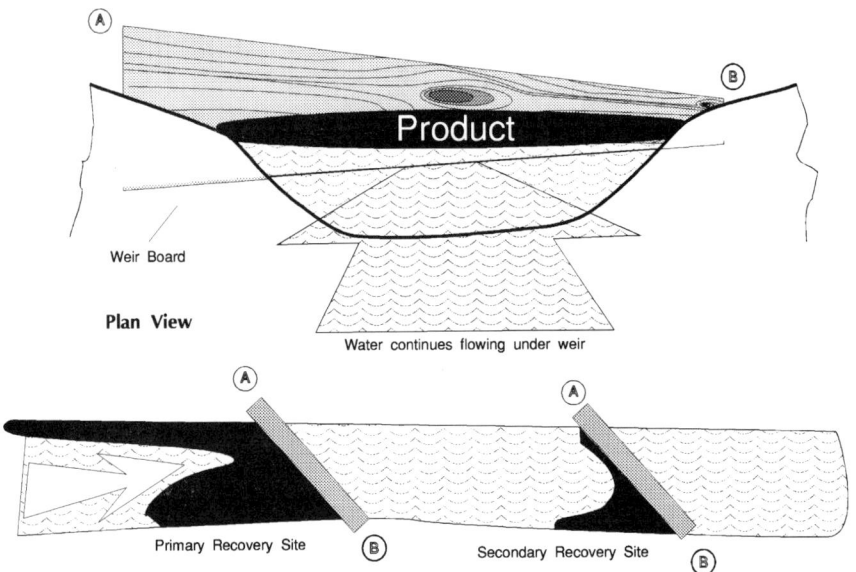

FIGURE 5.7.   Weir Construction

bed to raise the level of water. Once the level has been raised by approximately 12-in., one or more 4-in. pipes are placed in the dam to allow for water flow. Further construction then raises the water level to approximately 3 ft, as illustrated in Figure 5.8.

The level of water behind the dam will be controlled by blocking one or more of the pipes. Note the splash protection (such as plastic sheeting, rocks, etc.) on the downstream side to prevent erosion. The sheeting on the upstream side is to help reduce contamination of the soil used for dam construction and reduce the effect of dam porosity.

f. Culvert Weir
If a ditch or stream passes through a culvert, a very effective weir can be built across the culvert using plywood and other similar materials. Figure 5.9 illustrates this method.

g. Catchbasins and Sewers
Outside the confines of industrial facilities, many catchbasins and sewers feed to a municipal water treatment station or, worse, to open bodies of water. Many products will adversely effect the operation of treatment

Cross Section

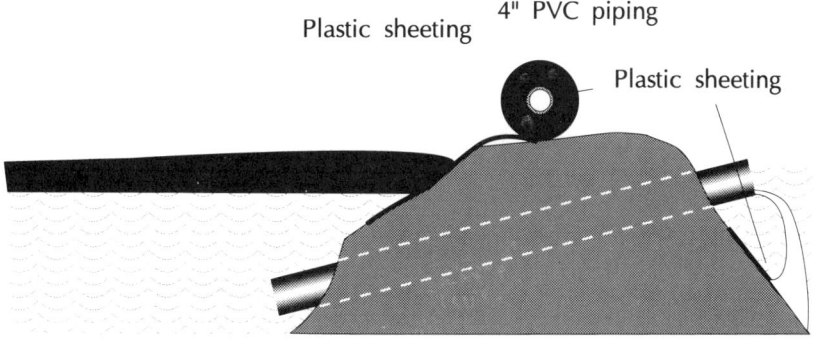

Plan View

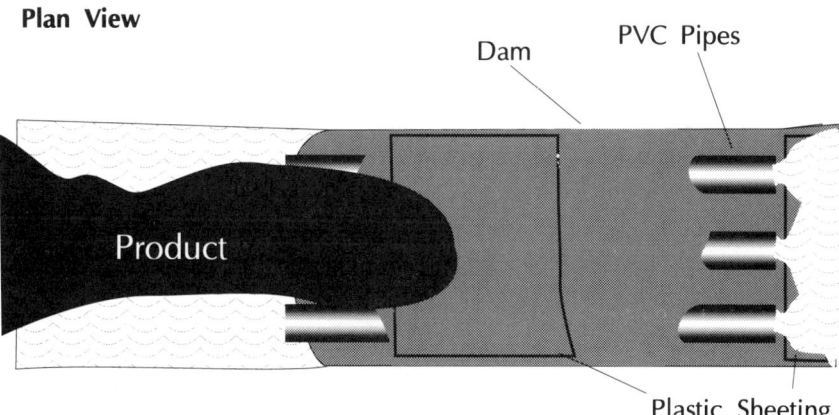

FIGURE 5.8. Dam Construction

facilities, and flammable products may lead to the accumulation of gases in the sewer system. Due to these additional complications associated with sewer invasion, it is necessary to cap catchbasins besides dyking around them.

Catchbasins may be capped by using patent devices or by using plastic sheeting (if the sheeting is resistant to the product) and sand. Whenever possible, dykes should be constructed in such a fashion as to encourage product movement away from catchbasins. Figure 5.10 illustrates this.

1B. Spills to Water

In considering spills to water, there are three distinct circumstances to be addressed:

Front View

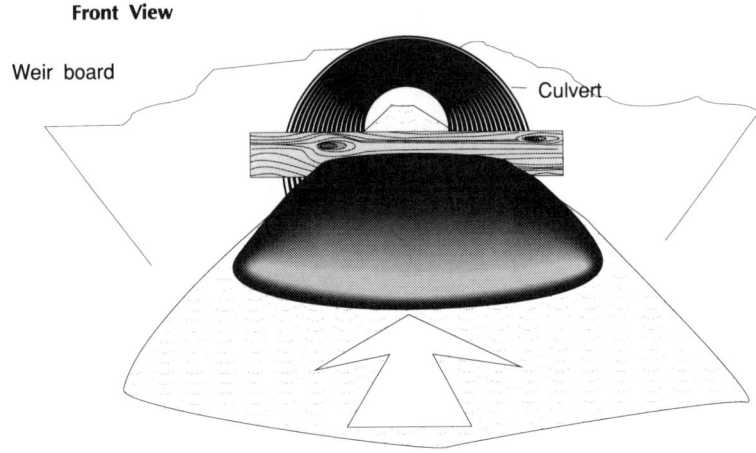

Side View

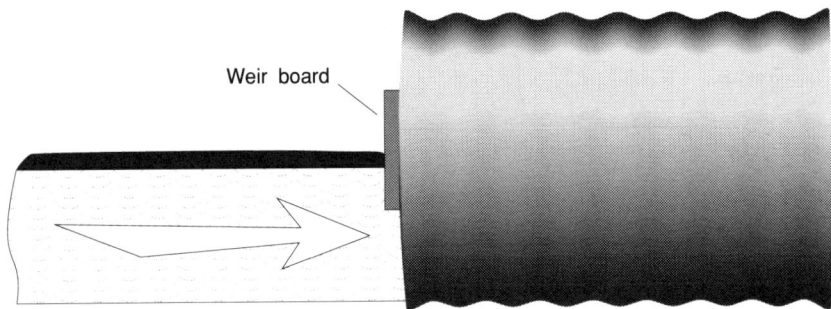

FIGURE 5.9. Culvert Weir

- Spills of water-soluble products
- Spills of nonsoluble products with a relative density greater than 1.0
- Spills of nonsoluble products with a relative density less than 1.0

Most of the equipment and techniques currently developed deal with spills of nonsoluble products with a relative density less than 1.0.

a. Soluble Products
Spills of soluble products require isolation and containment of the entire body of water affected. Rapid dilution of the product makes it difficult to assess where it has spread, especially in moving bodies of water. Detection is often achievable only by lengthy analytical processes, thus defeating the objective of response.

1. Containment   125

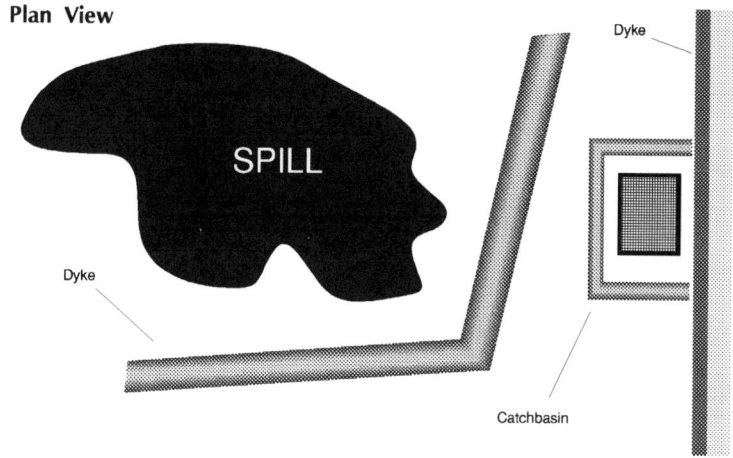

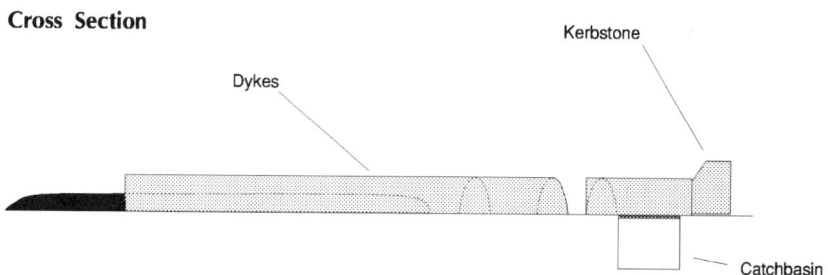

FIGURE 5.10. Catchbasin Protection

Unfortunately, most spills of this type are seldom resolved, unless they are in small, slow-moving bodies of water. The result of such incidents is often substantial contamination and interruption of municipal water supplies. Occasionally, with smaller product quantities in large bodies of water, the effect of dilution can be so marked as to effectively make containment and recovery impracticable.

The preceding accepted, there are techniques available for dealing with spills of this nature. One method is to construct a dam upstream of the spill area and another downstream. Pumps and piping are used to divert the stream flow around the area that has been contaminated; see Figure 5.11.

Another method is the construction of diversion channels. As with the preceding method, dams are constructed above and below the spill. A

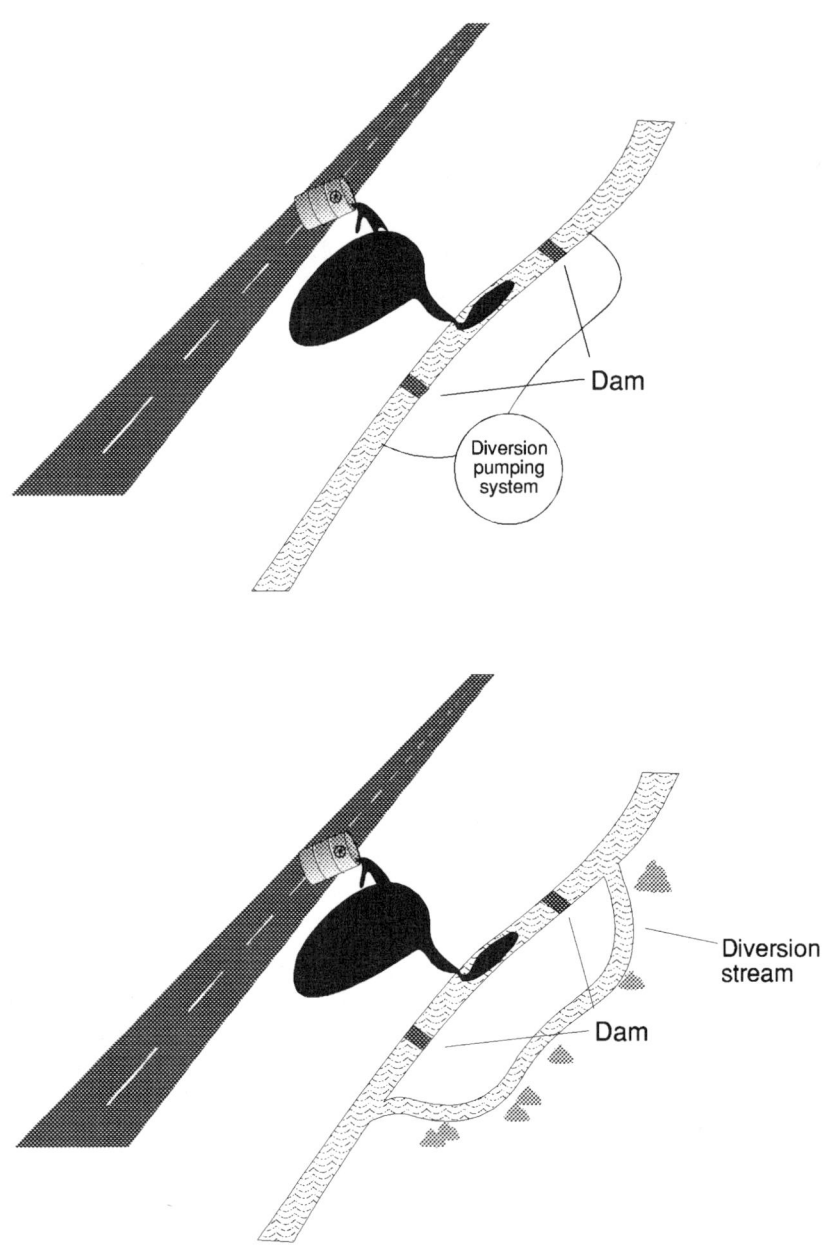

FIGURE 5.11. Stream Diversion

channel is constructed to allow clean water to bypass the section of contaminated water.

The effective implementation of either of these methods greatly depends on the watercourse characteristics, spill location, availability of large-scale equipment, and speed of implementation. As mentioned before, identification of the area that is contaminated is critical.

b. Nonsoluble Products—Density Greater than Water

Several products are nonsoluble in water and are more dense than water, for example, some PCB oils and dry-cleaning fluids. The primary problem associated with such products is that they are difficult to observe in a spill situation: Their location often requires the use of divers or sounding equipment.

The bottoms of most waterways are irregular in shape and act as a containment barrier in themselves. The distance a product travels will depend largely on a combination of water velocity and bottom makeup.

Where a spilled product remains immobile on the bottom, it is acceptable to proceed directly to the recovery stage. Where the product is moving downstream, efforts should be made to limit its movement. An effective way is to deposit heavy materials in the stream bottom of the waterway, therefore creating a damming effect.

Alternatively, instead of depositing materials to form a dam, it may be possible, especially in shallower waters, to construct a trench along the bottom. These two techniques are illustrated in Figure 5.12.

c. Nonsoluble Products—Density Less than Water

As mentioned in the opening comments of this section; most equipment and techniques used in spill management are aimed at nonsoluble products that are less dense than water. Before reviewing the containment methods available, it is worth considering the ways in which floating products break up. This is illustrated in Figure 5.13.

While it is beyond the scope of this book to discuss these processes in detail, it is noteworthy that product will tend to dissipate when it is spilled. Many of the techniques discussed assume that the product remains intact. Effects such as sinkage and emulsification may affect the ability to contain and recover the product.

The following methods and apparatus are available for product containment.

Mechanical Boom
The mechanical boom is by far the most effective and often used method for product diversion and containment. There are many different kinds of

128   Spill Control

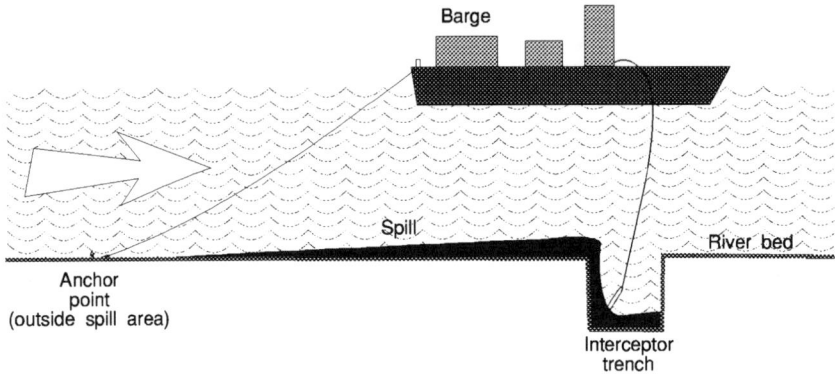

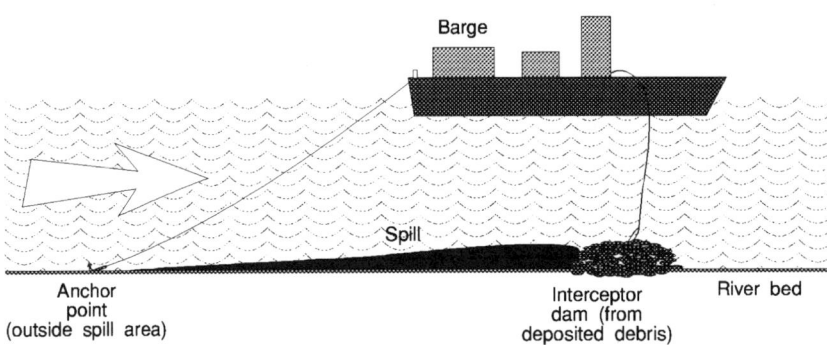

FIGURE 5.12.   Containment of Dense Products

patented booms available on the market, including some that can be deflated and rolled up into a drum. The most-often encountered boom is illustrated in Figure 5.14. This type of boom has an upper freeboard section that incorporates polyurethane foam air chambers to provide flotation and surface containment. A lower section, called the skirt, prevents subsurface loss and product carryover. The skirt is a fabric material ballasted by chain, lead weight, or cable. Tension members in the boom assume the load from water pressure acting on the boom and help to maintain the boom in an upright position. Booms are available in a variety of sizes and materials, from as small as 4 in. to approximately 6 ft overall height.

Containment with booms is effective in currents up to 1.5 knots. Above this velocity, containment is difficult. When the current velocity exceeds

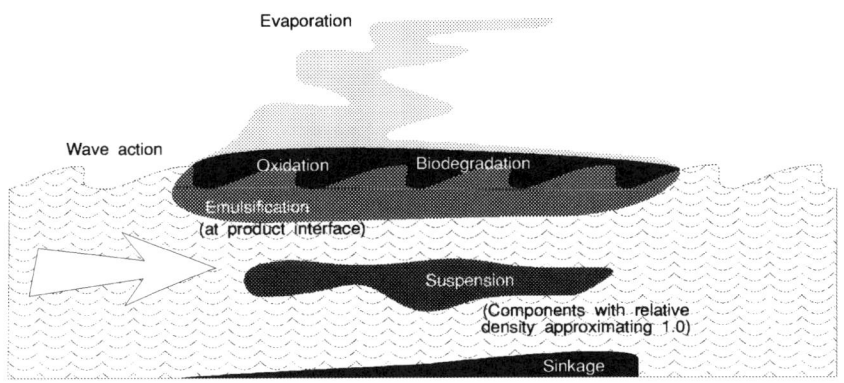

FIGURE 5.13.   Product Breakup

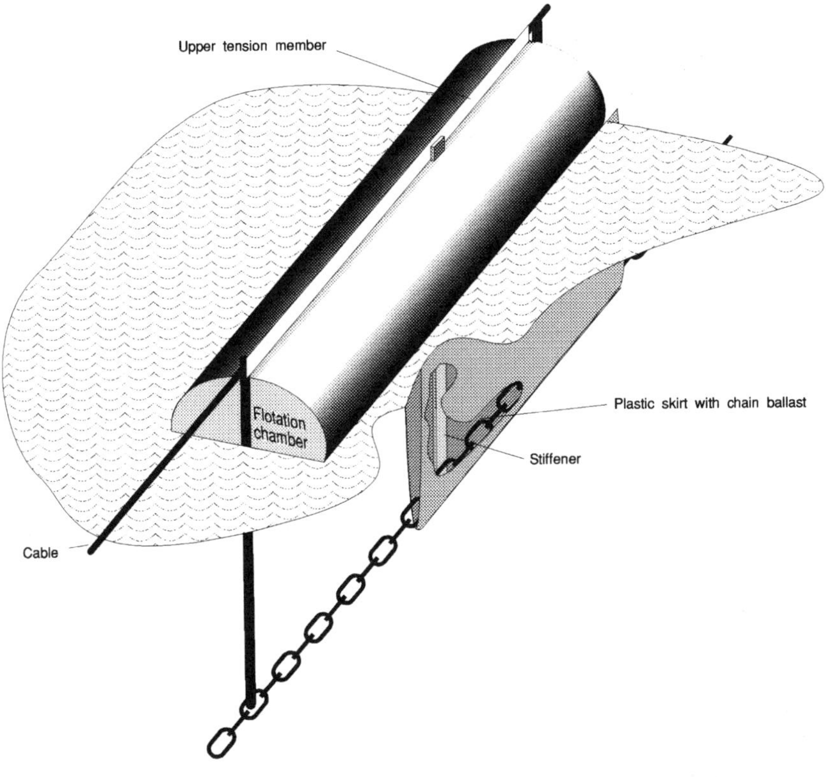

FIGURE 5.14.   Mechanical Boom

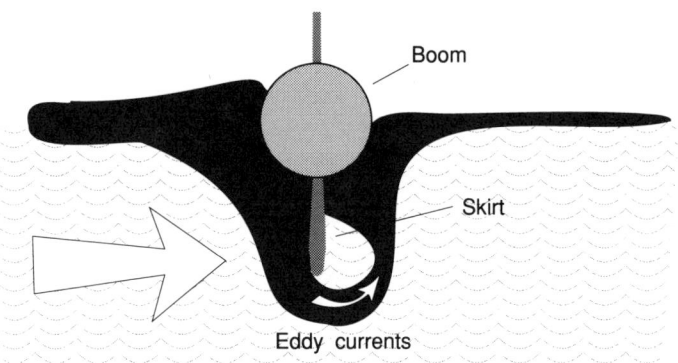

FIGURE 5.15.   Product Entrainment

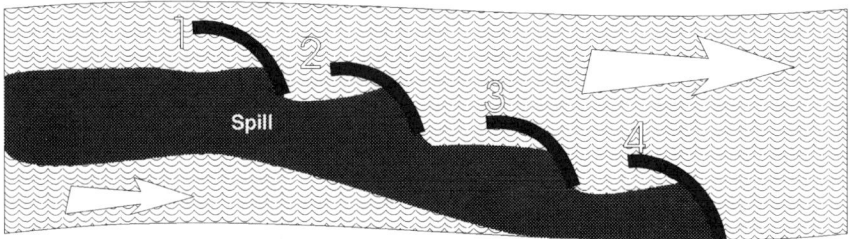

FIGURE 5.16.   Boom Cascade Deployment

0.75 knots, the floating product may begin to entrain beneath the boom or carry over. This condition is illustrated in Figure 5.15. This problem is addressed by angling the boom to the direction of current flow, thus reducing the relative velocity at the boom face. Once the boom is angled, it will cause the product to flow down the face of the boom. This makes necessary the use of a longer boom or, more correctly, the deployment of several booms in what is known as a cascade. This technique is illustrated in Figure 5.16. The boom is deployed by using a system of ropes and anchors to achieve the desired angle to the current.

Boom Alternates
For small spills in ditches and creeks where a mechanical boom is not available, there are several possible alternatives. These include straw booms, sorbent booms, and log combinations, and are illustrated in Figure 5.17.

Straw Boom

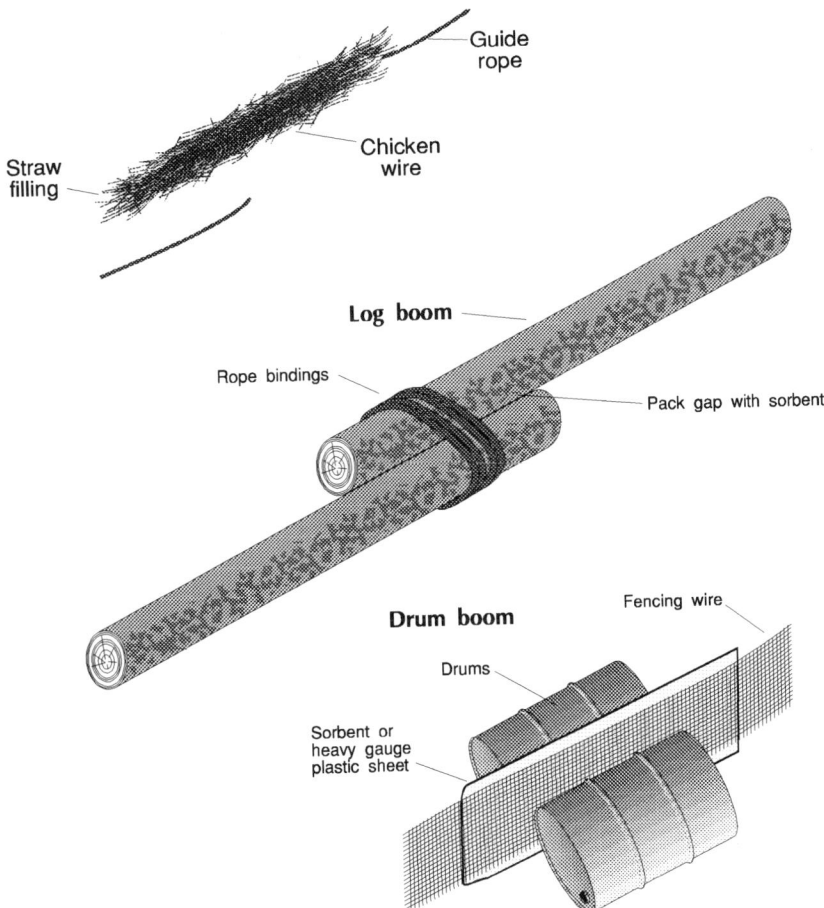

FIGURE 5.17. Mechanical Boom Alternatives

2. RECOVERY

The recovery of soluble products from water is largely a matter of mass removal by pumping. Recovery of dense, nonsoluble products is normally achieved by dredge and vacuum suction. The primary concern in this area is the recovery of nonsoluble products from water, and it is this that merits the most attention.

Generally, techniques are designed to recover the largest possible quantity of oil and the smallest quantity of water. This objective is achieved by one of several methods, which are addressed in this section. Once spilled product is recovered, it must be placed in suitable temporary storage, such as portable tanks or vacuum trucks.

Besides the unit used for separation, it is necessary to use a prime mover. These are normally centrifugal, gear, or diaphragm-type pumps, depending on the product. Vacuum trucks provide a satisfactory alternative, but they are often hard to find and have difficulty accessing remote areas.

Recovery equipment falls into one of three main categories:

- Skimmers
- Mops
- Sorbents

2A. Skimmers

A large variety of skimmers are available for spill recovery. The most commonly encountered are weir-type units. Within each category there is a proliferation of types, the main differentiation being volumetric capacity. Table 5.2 discusses the advantages and disadvantages of the various types of skimmers.

TABLE 5.2   Oil Skimmer Features

Type	Advantages	Disadvantages
Weir	Handles most products, including chemicals and heavy oils. One moving part. Simple maintenance. Relatively cheap.	Tendency to skim too deep on light products. Needs constant flow type pump. May clog with debris.
Rotating Disc	Very high product/water ratio. Prevents debris reaching pumping apparatus.	Poor at recovering heavy oils. Channels clog. High maintenance. Discs may react to certain chemicals.
Oleophilic Rope	Good for heavy oils. Acts as barrier to product movement. Prevents debris reaching pumping apparatus.	Rope is very difficult to decontaminate. Oleophilic properties deteriorate during use.
Suction		Channels clog easily.

Hydro - Adjustable Weir

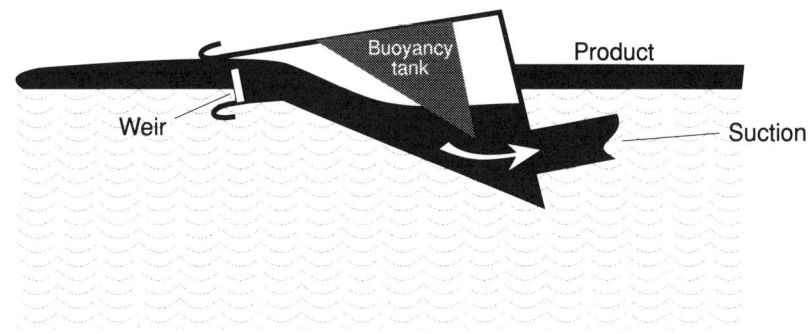

Saucer Weir

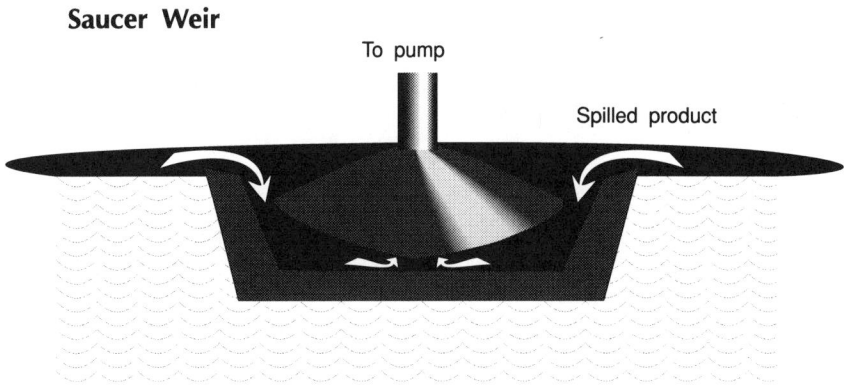

FIGURE 5.18. Weir-Type Skimmers

a. Weir Skimmers

Weir skimmers consist of a recovery tank fitted with a floating weir chamber. This arrangement allows floating oil to pass through the unit while preventing the water below from passing over the weir mouth. Increased pumping rates cause the unit to cantilever and lower the weir face deeper into the product and, eventually, water. Figure 5.18 illustrates a typical weir skimmer.

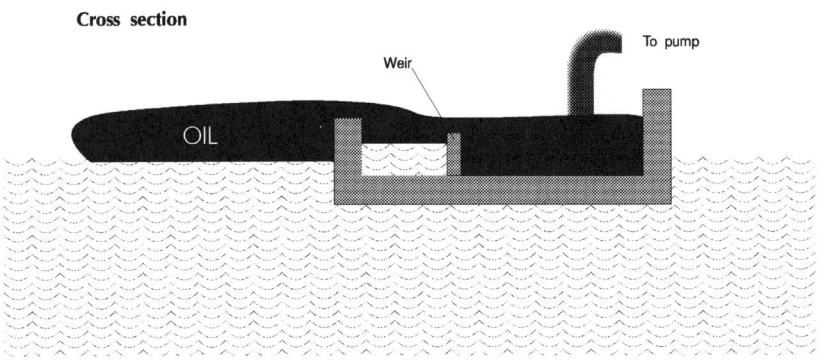

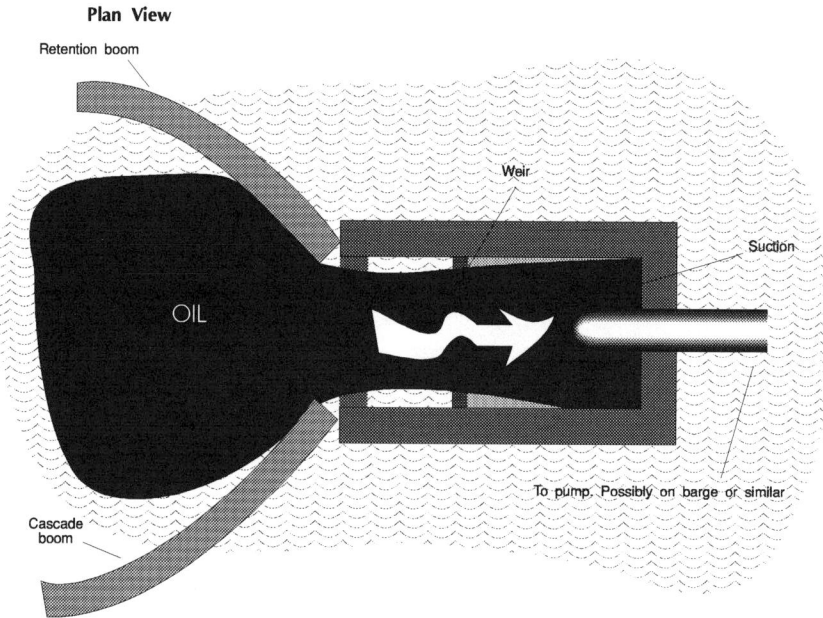

FIGURE 5.19. Large Weir Skimmers

As with most types of recovery equipment, weir skimmers come in a variety of shapes and sizes. The smallest units resemble a vacuum cleaner head, are hand held, and are guided by an extension piece. The largest units weigh more than 200 kg and are designed to fit into a mechanical boom pattern. Figure 5.19 illustrates the deployment of a unit.

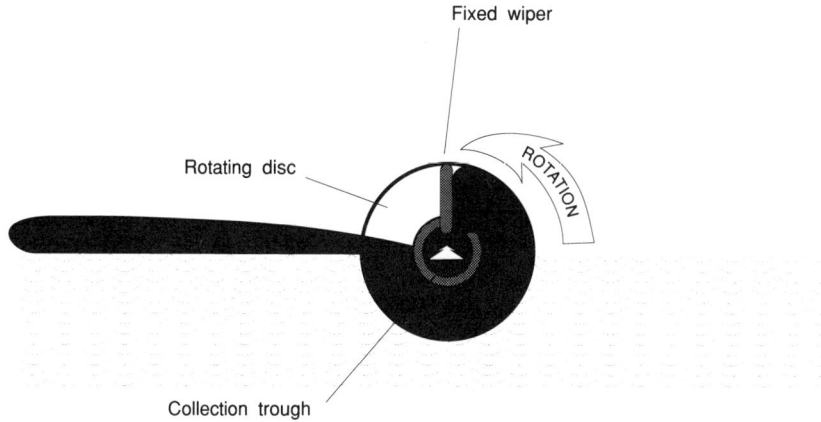

FIGURE 5.20. Oleophilic Disk Skimmer

b. Disk Skimmers

Rotating disk skimmers use an oleophilic ("oil-loving") disk that rotates through an oil/water mixture. Oleophilic materials possess the properties of an attraction for oil coupled with the ability to repel water. Therefore, as the disk passes through an oil/water mixture, it will collect oil while recovering very little water. Figure 5.20 shows the construction of such a unit.

After the disk passes through the liquid, the residual oil is removed from the disk surface by means of a fixed wiper. The wiper is concave in shape and channels the recovered oil to a collection trough, which is at the center of the unit. From the collection trough, the recovered oil is pumped out by means of a built-in discharge pump.

The smaller versions of these units are electrically driven and require a portable generator; the voltage used depends on the unit, but typically is 12 or 24 V d.c. When using such units with flammable products, be careful to ensure that the unit is intrinsically safe.

The larger units use a hydraulic power pack for primary drive and are fitted with hydraulic on-board motors for disk rotation and pumping out recovered product. The principal advantage of this arrangement is that the units offer a high degree of safety when handling flammable products. The prime mover can be located in a position away from flammable or potentially flammable vapors.

As mentioned in Figure 5.19, when selecting any of these units, be careful to ensure that the disk material is compatible with the product to be handled.

136   Spill Control

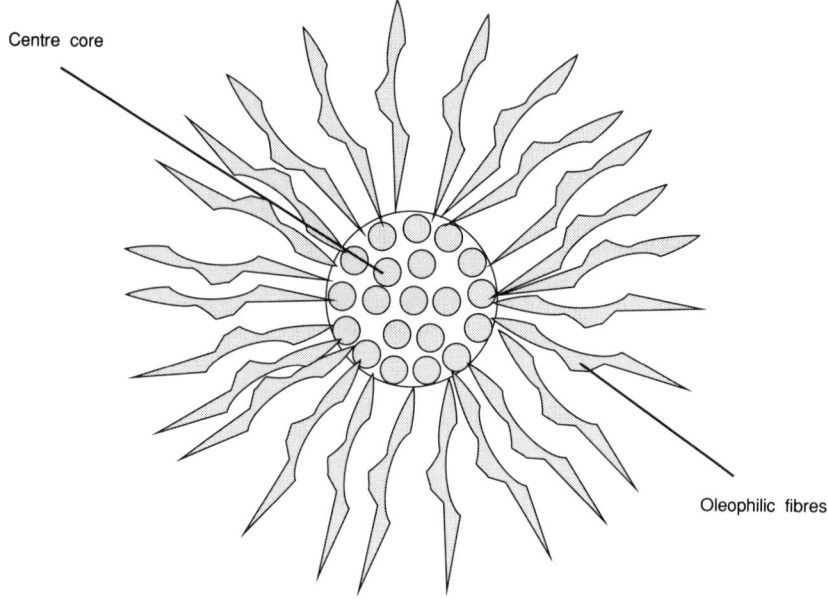

FIGURE 5.21.   Oleophilic Rope

c. Rope Skimmers
Like the disk-type skimmers, rope skimmers rely on the oleophilic properties of certain materials to recover oil while repelling water. Rope skimmers consist essentially of a "hairy" rope that passes through a squeeze-roller arrangement. The rope is a normally spun center with strands of the material woven through it. Figure 5.21 shows the rope construction.

The rope is deployed so that it passes through the spilled material, usually a heavy oil such as bunker. Rope recovery is effected by means of a squeeze-roller arrangement that is positioned over some type of collection unit. Figure 5.22A shows a typical deployment. Figure 5.22B illustrates how a rope can be deployed using two tail blocks around a spill area. The principal advantage of this arrangement is that the rope acts as a barrier or "boom" and restricts product movement to some extent. It should be noted that oleophilic rope is not a substitute for a mechanical boom. Care should be taken to ensure that the area between the shoreline and the recovery unit is protected from recovered product that will fall from the rope as it traverses this distance.

d. Suction Skimmers
Suction-type skimmers work on the principle of taking a suction source and "spreading" it evenly over the surface. The most commonly encountered of those are the "Manta"-type units, as illustrated in Figure 5.23.

2. Recovery   137

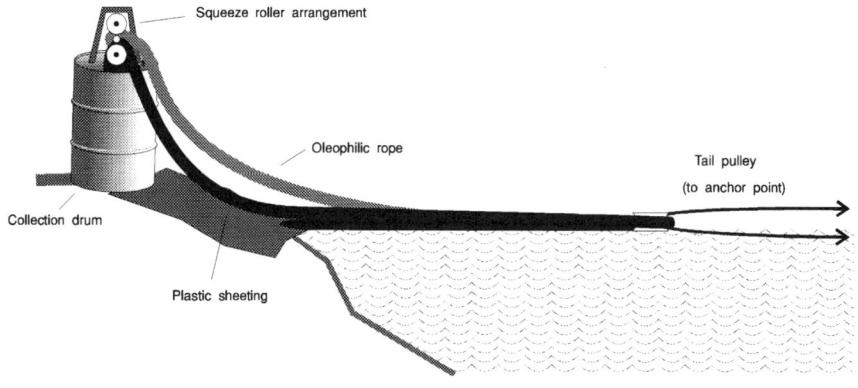

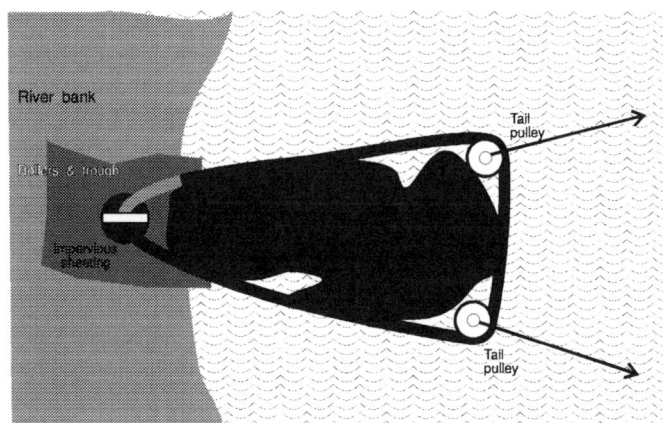

FIGURE 5.22.   Rope Deployment

The unit is designed to float on water. Given the ratio of hose diameter to suction area, it can be appreciated that the face of the unit is extremely narrow. This ensures that only the top of the liquid surface is recovered, which will consist mainly of spilled product.

The narrow suction face is the cause of the main problem with this type of unit, because it will readily clog in the presence of debris or viscous products. In situations such as a gasoline spill in a marina, these units are ideal. However, when deployed on a heavy product spill in a large body of moving water, their effectiveness is limited. These units are best used with a self-priming, constant-velocity prime mover, such as a vacuum truck or diaphragm pump. They tend to entrain significant volumes of air, especially when operated by inexperienced personnel.

138   Spill Control

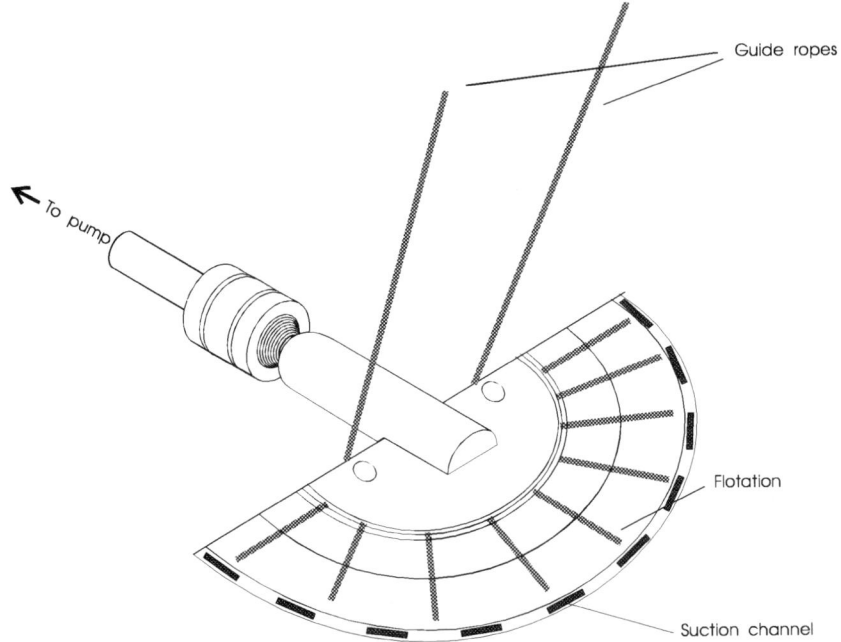

FIGURE 5.23.   "Manta"-Type Suction Skimmer

2B. Sorbents

Numerous types of sorbents are available for spill recovery. The precise mechanics of how each type of sorbent works are not discussed here. There are two major categories of sorbent, *ad*sorbent and *ab*sorbent; their characteristics dictate the manner in which they are deployed.

Examples of adsorbents are soil and sand. They are used by mixing with a spilled product until the resulting mixture is recoverable by some manual method, i.e., a rake, shovel, or fork. Some adsorbents will float on water and can be used for water-based spills; treated peat moss is one example.

Examples of absorbents are diapers, straw, and most commercially manufactured sorbent products. These materials become wetted with the product when in contact with it and continue to absorb it until they are saturated. Some absorbents are oleophilic in nature; i.e., they will absorb oil but not water. Others have been developed for the recovery of very specific materials. When absorbents are to be used for recovery of corrosive materials, ensure that they are sufficiently resistant to the corrosive effects of the product.

Synthetic absorbents come in several configurations, including pillows, pads, rolls, and booms. In many cases they can be compressed and the recovered contaminants rung out from them. Once this is accomplished, it may be possible to reuse the absorbent material on the spill. When any sorbent product, regardless of the material recovered, is set aside for disposal, the risk of spontaneous combustion is extremely high, and precautions should be observed.

2C. Temporary Storage

This chapter has reviewed the techniques of product containment and recovery. As a final topic, it is worthwhile to review the temporary storage facilities that may be available or developed at a site. These methods are reviewed in ascending order of size.

a. Drums

Commercial drums make effective temporary storage at spill sites, especially where small volumes of product and debris are recovered. As with all circumstances, ensure that the drums to be used for storage are compatible with the recovered product.

To place the recovered product and materials in the drum it will be necessary to remove the drum lid, or shear the drum top from the body. Where the latter action is used, the drum is no longer suitable for transporting recovered materials. Also, with products that vaporize, the drum will not contain vapors and it must be stored at a well-ventilated site.

If a drum has been altered as described previously, the contents must be transferred to an intact drum or the existing drum may be overpacked or transferred to a lugger box.

b. Lugger Boxes

Lugger boxes are readily available from most waste management companies. These units can be in either closed or open-top configurations. The main advantage of lugger boxes is the volumetric capacity; a sometimes bothersome disadvantage is siting the units close enough to a recovery site. Where this is a problem, interim units, such as drums or small dumpsters, should be used. Lugger boxes are frequently used to transport drums that have been physically altered or damaged.

c. Portable Tanks

There are many types of portable tank. They vary in size from 1895 to 37,900 liters (500 to 10,000 gallons) in capacity. Their construction is essentially of two types: either the self-supporting bladder type (limited to the smaller sizes) or, more typically, framework and liner.

There are a few precautions to be taken when using these units:

- Never exceed the tank's rated capacity.
- Ensure that the liner material is compatible with the product to be recovered.
- Remove stones, sticks, and any other protruberences from the area in which the tank is to be sited to avoid the risk of tank punctures.
- If possible, site the tank well away from or downhill of waterways.
- Keep one person at the tank at all times to monitor the liquid level.

d. Transport Units

Where large quantities of liquids are involved, vacuum trucks, tank trucks, and rail cars all may have a use. Vacuum trucks are often not as readily available as other units. It is a good strategy to use one or two vacuum trucks and allow them to transfer into tank trucks or rail cars.

e. Temporary Pits

In some circumstances, there may be insufficient capacity available for storage in the previous units discussed. It may be impractical, for example, to store large volumes of contaminated material and heavy oils in lugger boxes.

In such circumstance it may be possible to obtain approval from environmental authorities for the construction of temporary pits and ponds. Criteria used in deciding on location are soil porosity, level of groundwater, proximity of surface waters, and likelihood of precipitation. Wherever possible, natural depressions rather than excavations should be used.

Before depositing debris, the pit should be lined with clay or industrial-grade plastic sheeting.

3. DRUMS AND CYLINDERS

Many incidents involve assortments of cylinders, drums, and carboys. This section offers guidance on dealing with these units and effecting temporary repairs at the incident site.

3A. Drums

Two main types of drums are in common use for the transportation of liquids: lined and unlined. Unlined drums are used for products such as

3. Drums and Cylinders   141

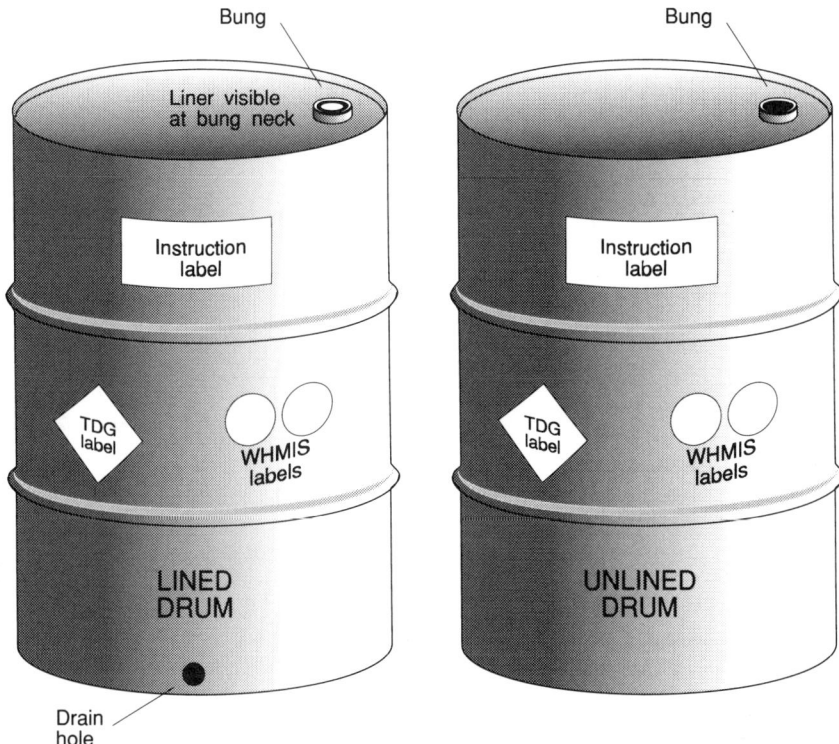

FIGURE 5.24.   Drum Features

gasoline, diesel, oil, kerosine, etc. Lined drums are used either for products that will cause degradation of the drum itself or for products that will suffer loss of quality upon exposure to steel.

While the lining material varies according to the drum contents, the general arrangement of lined drums is in essence the same. Figure 5.24 illustrates the typical appearance of the two types of drum.

All drums that contain hazardous materials require safety marks to be applied. These marks are dictated by several U.S. and Canadian statutes, and are designed to indicate the hazardous nature of the drum contents. A typical example, gasoline, would have a TDG/49 CFR flammable liquid label, a WHMIS/Right to Know flammable and combustible label, with a label indicating emergency procedures.

Upon encountering an incident involving drums or carboys, the following procedures should be adopted.

- Unless the exact contents of all containers are known, select the highest level of protection (Level A).

142   Spill Control

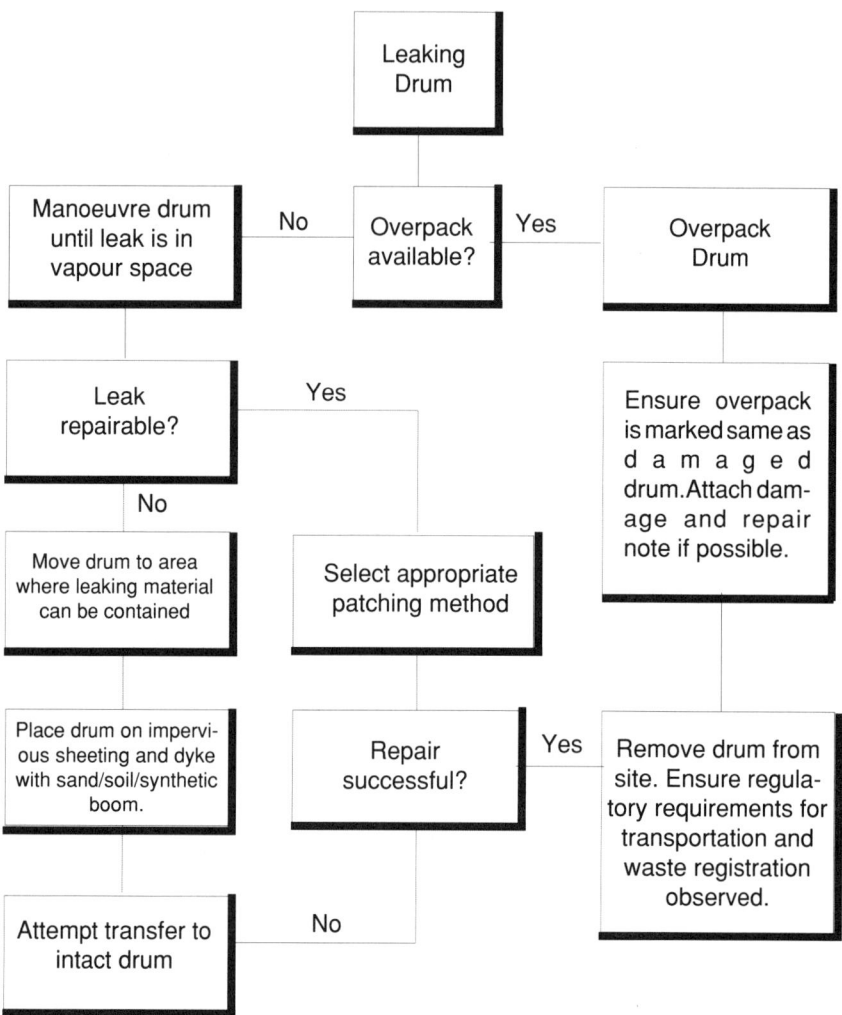

FIGURE 5.25.   Drum-Handling Decision Process

- Approach the spill site and determine clearly the contents of each drum; note which drums are leaking. If the shipping document is available, try to compare it to the drums found.
- Consult appropriate technical data (MSDS and spill guidance forms) to assess the potential for reactivity.
- Reenter the site and stabilize any leaking units by repositioning if possi-

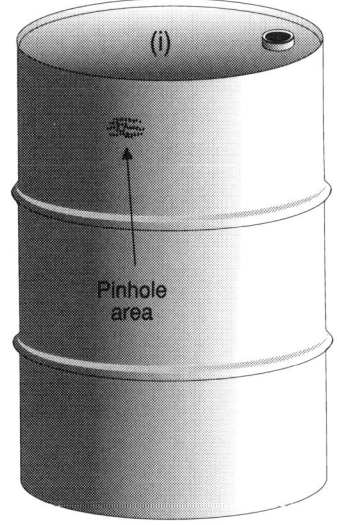
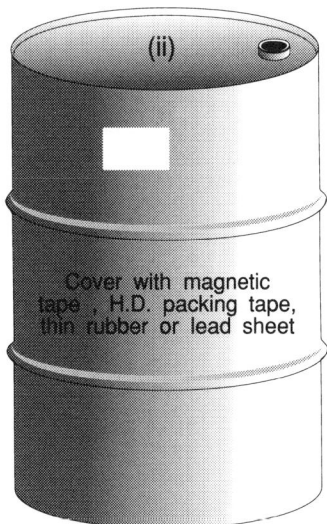

FIGURE 5.26. Pinhole Leaks

144   Spill Control

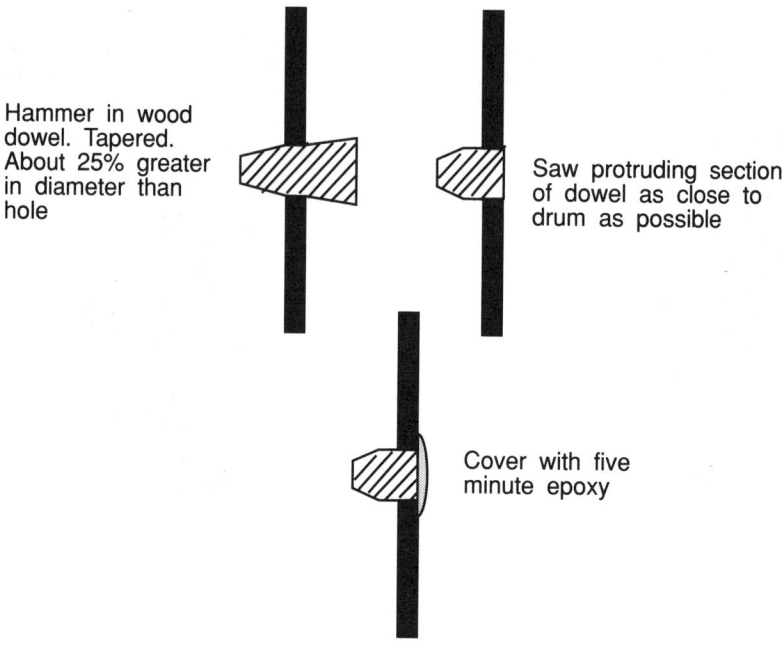

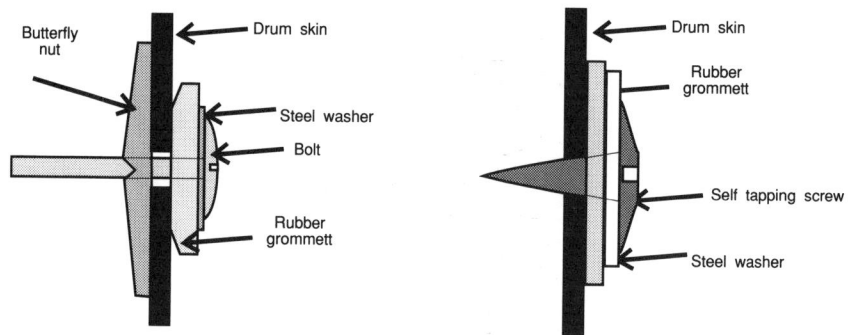

FIGURE 5.27.   Small Holes

ble. That is, if a drum is holed and the hole is at ground level, attempt to rotate it until the hole is adjacent to the vapor space.

The next stage involves carrying out temporary repairs to damaged drums and overpacking. Often there will be insufficient overpack drums available to allow overpacking of all units, so all temporary repairs should

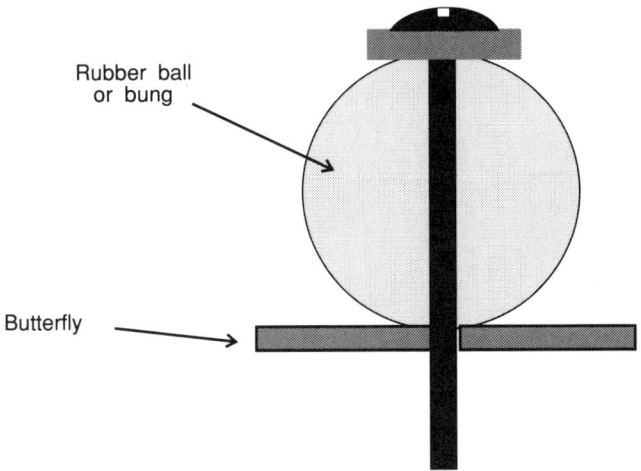
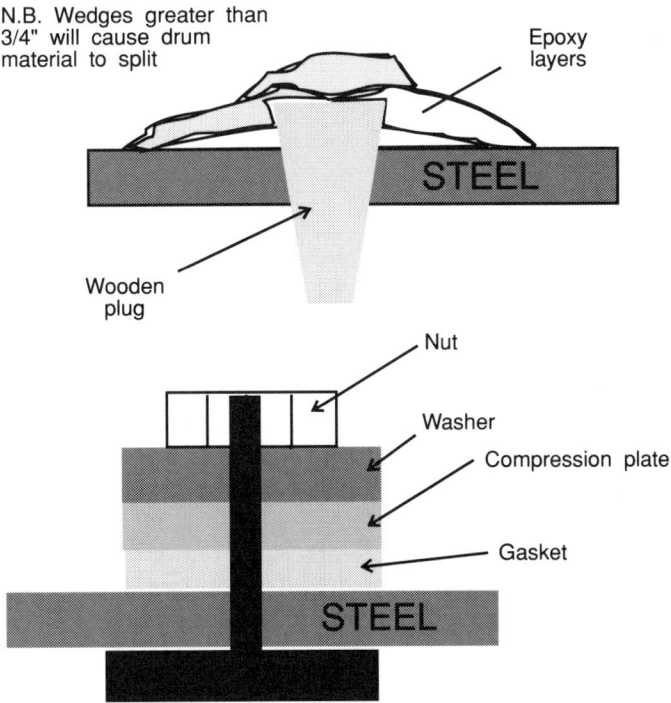

FIGURE 5.28.   Large Holes

145

TABLE 5.3   Drum Patch Kit. Suggested Equipment List

Tools	Fittings
Crowbar, 24″	Balls, plumber, rubbers, assorted
Hacksaw, 14″	Clamps, "G" type, 3″
Hacksaw, junior	Clamps, "G" type, 7″
Hammer, brass 2 lb.	Dowels, assorted 1/4″ to 3/4″ diameter
Hammer, steel 2 lb.	Screws, self tapping, assorted
Knife, carpet	Screws, with ring nuts and grommets, assorted
Knife, swiss army	
Pliers, long nose 6″	
Pliers, snub nose 6″	
Screwdrivers, universal 6″	
Wrench, adjustable 8″	
Wrench, bung universal	
Wrench, crescent 12″	
Wrench, pipe 12″	
Miscellaneous	
Epoxy adhesive kit, 5 minutes	
Flashlight	
NIOSH Pocket Guide	
Rubbing alcohol	
Sandpaper	
Tape, aluminum	
Tape, magnetic	
Tape, teflon	
Wipes	
Wire wool	

be designed to permit safe transportation to a site where the drum contents can be handled.

Another alternative to consider before patching is transferring the contents of a damaged drum into an intact unit, but again, this will depend on the availability of clean drums.

The flowchart in Figure 5.25 summarizes the decision process discussed so far.

Where drum repair is to be attempted, use one or more of the techniques shown in Figures 5.26, 5.27, and 5.28. There are several commercially available kits designed specifically for these techniques. A typical kit is shown is Table 5.3.

Additionally, various types of chemical patch kits are available. These either require the mixing of two chemicals to form a malleable material that sets hard, or a single malleable material that sets hard upon exposure to air. Both types have their uses and should be included in a drum patch kit.

When a hole is to be repaired with self-tapping screws, wedges, dowels, or other such equipment, the following procedures should be observed:

- Select an appropriate arrangement for the size of hole to be repaired.
- Ensure that all materials are resistant to degradation by the product inside the drum.
- Where practicable, ensure that the leak is in the vapor space.
- Clear the area surrounding the hole by sanding, rubbing with wire wool, and finally wiping down with alcohol.
- Insert the sealing unit and drive through the dowel or wedge.
- If a dowel is used, saw it flush with the drum surface; with a wedge, leave about 1/4 in. (6 mm) of the wedge protruding; tighten screw-type units.
- Rewipe the area with alcohol and then cover the seal interface with 5-minute epoxy.
- When a wedge is used, apply a second layer of epoxy after 5 minutes.

3B. Cylinders

In most instances, especially during the early stages of an incident, the only precaution that can be taken with cylinders is to isolate them so that the leak is located in the vapor phase. Occasionally, simple things such as closing valves that have sprung may prove effective.

Attempts should be made to place the cylinder in a safe, well-ventilated area. If appropriate (i.e., for ammonia), use a water fog to knock down the escaping gases.

There are several patent units available for dealing with specific cylinder types, such as the series developed by the Chlorine Institute. However, such units are specialized pieces of equipment that are not readily available. Many of these are held by shippers of the products and local TEAP/CHEMNET teams.

6

Air Monitoring

One of the most significant risks of general exposure is that of contact with airborne contaminants. At any incident site it is vital that unprotected personnel do not have contact with contaminants at concentrations above the published exposure limits. To ensure that contamination does not occur, it is necessary to monitor the atmosphere at any incident site. This chapter discusses protocols of air monitoring and the instruments used.

There are several objectives of air monitoring, including the following.

Ensuring safety of personnel. Response teams must ensure that they and others are not exposed to any concentration above TLV—TWA. One method of avoiding this is to approach all incidents from an upwind position. This may not always be possible, because of topo/geographical considerations, a no-wind condition, or product migration in a direction opposite to the wind. No matter what the circumstances, any area designated for general access should be constantly monitored.

Confirmation/downstaging levels of protection. While initial approaches are always made utilizing a high level of protection, at some stage of an incident a response team hopes to lower the level of protection required for various areas on the site. Once the concentration of airborne contaminants is characterized, it may be possible to reduce the level of protection required. This is particularly so for areas such as decontamination stations, equipment cleaning, and other support activities. Working in high levels of protection is arduous, time consuming, and poses the highest risk for accidents. In order to "step down" from Level B to Level C, it is necessary to implement an air monitoring protocol that ensured wearers did not compromise the maximum rating of their air-purifying equipment (i.e., with respect to IDLH, APFs, and oxygen concentration).

Confirm incident control zones. In the initial stages of an incident, control zones will be established with a margin for error in initial assessment. They will extend considerable distances beyond what is required. This is an acceptable approach. By establishing an air monitoring protocol, however, it may be possible to considerably reduce the size of incident control zones. In making these decisions on the basis of values monitored, other factors such as minimum safe distances must not be compromised.

Monitor exposures. Unfortunately, exposures do occur at incident sites. If an adequate air monitoring protocol has been used during an incident, it should be possible to qualify the potential health threats to any personnel who have been exposed. Without proper monitoring it will be impossible to deduce any reasonable value of exposure.

Characterize the site and contaminant spread. By monitoring and mapping concentrations at various points, it is possible to characterize the spread of contaminants and the development of plumes. Extrapolation of data obtained will aid in deciding what, if any, risk is posed to personnel in downwind positions. An additional benefit of air monitoring, for products that have associated vapors, is that product presence may be indicated by the detection of its vapors.

1. VALUES MONITORED

The exact values monitored at various points depends on the subjective assessment of incident personnel. Figure 4.27 offers an example of the level of values to be assessed over a period of time. In the initial stages of a monitoring protocol, the primary effort is to rapidly characterize the threat presented and obtain some indication of vapor migration. Remember, all responses in the early stages will be in higher levels of protection, including air monitoring activities. The first set of values obtained should be as shown in Figure 6.1.

To determine the vertical dispersion of contaminants, samples should be taken at approximately 2 m (6 ft), and 1 m (3 ft) above ground level, and at ground level. If any point gives an unacceptably high reading, the monitoring team should move away from the spill site until an acceptable value is obtained. For example, if position 3 in Figure 6.1 gave a reading of 50 percent LEL on a combustible gas indicator, the team should retreat downwind until a reading that has been deemed safe is obtained. Generally, this reading would be 20 percent LEL. The values monitored at this stage will depend on factors such as any known threat posed by the spill (i.e.,

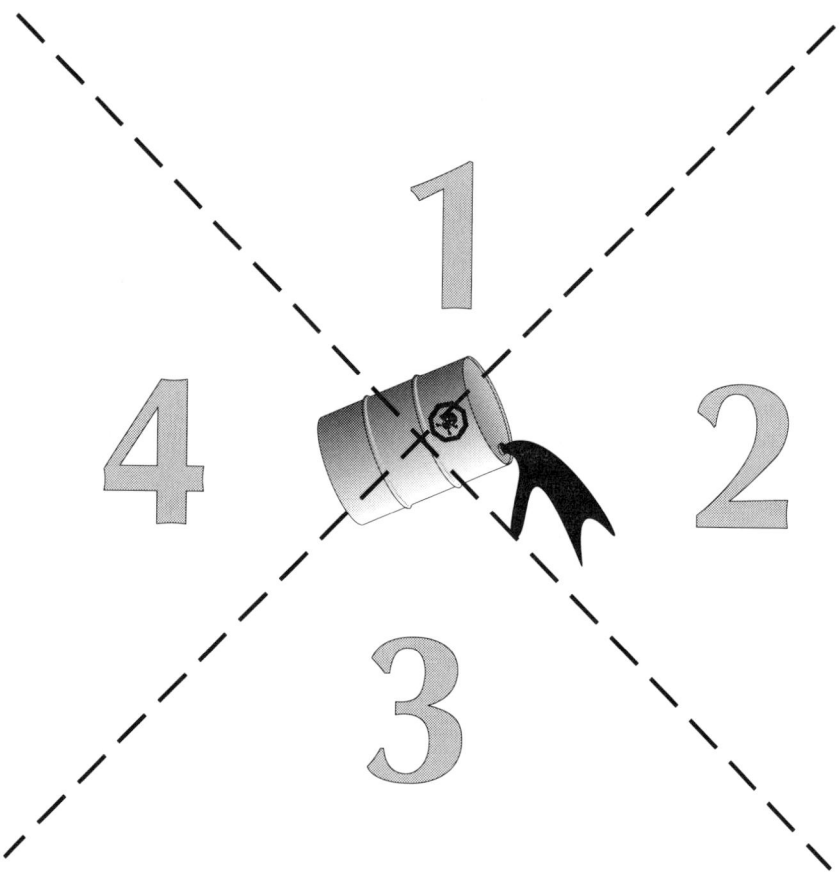

FIGURE 6.1. Initial Monitoring Pattern

flammability, corrosivity, or toxicity) and the types of instruments available.

There are several other documented initial monitoring patterns, but the one presented in Figure 6.1 will allow response teams to rapidly assess the degree of lateral dispersion (from positions 2, 3, and 4) and obtain an upwind safety check from position 1.

The discussion so far has assumed that something is known of the hazardous nature of the product involved in the incident. In some situations a response team may be unable to determine the product involved. In this case, the primary approach will be to categorize the product based on air monitoring results. The first assessment in air monitoring is for risk of combustion and oxygen deficiency. The site must be characterized from these perspectives first. Once these two risks are evaluated, the team will next address the risk of lower concentration and toxins.

2. SAFETY IN AIR MONITORING

Air monitoring teams are no less susceptible to the adverse effects of hazardous materials than other personnel working at an incident site. All premises of safe approach, such as use of the buddy system, constant visual contact, prepared backup, personal protective equipment, and decontamination protocols must be in place prior to sending an air monitoring team in to approach an initial exclusion zone.

3. AIR MONITORING INSTRUMENTS

The range of air monitoring instruments available to responders spans from rudimentary combustible gas indicators through to portable gas chromatographs. In practice the critical stages of an incident are normally managed using instrumentation at the rudimentary end of the scale. Typically, response personnel will have available to them the following portable instruments;

- Oxygen analyzer
- Combustible gas indicator
- Colorimetric detection tubes
- Photo-ionization detector
- Product-specific units

Other, more discriminatory, instruments can be utilized in the later stages of an incident, especially in determining soil cleanup criteria, personnel monitoring, etc.

3A. Oxygen Analyzers

In many response situations there will be a safe level of oxygen present, i.e., above 19.5 percent. However, it is essential to monitor oxygen concentrations prior to and during operations at a spill site. As with other situations, some judgment calls are necessary; it could be deemed a misuse of time to check for the presence of oxygen in a situation involving 40 liters (10 gallons) of furnace oil spilled on open ground. Whenever there is any reason to suspect an oxygen-deficient atmosphere, monitoring should be initiated. As mentioned previously, oxygen-deficient air is generally heavier than normal air and will be found in lower-lying areas.

Oxygen analyzers work on the general principle of inducing a small

152   Air Monitoring

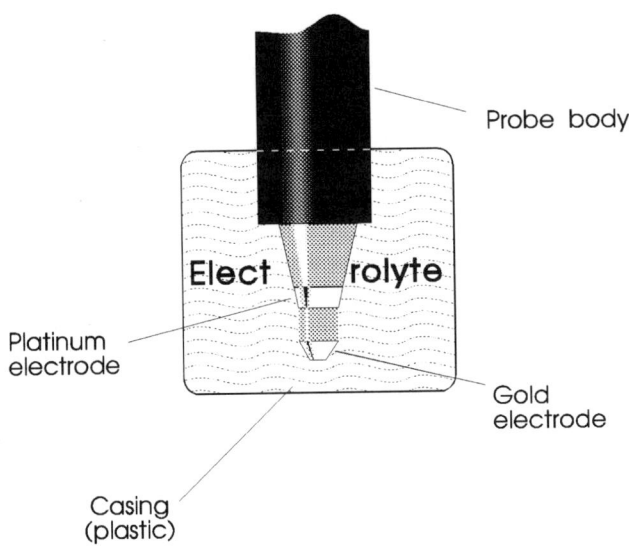

FIGURE 6.2.   Oxygen Diffuser Head

electrical current into a detector, amplifying the current, and displaying it as a direct-reading output. Two types of oxygen sensors are in common use, those with paramagnetic sensors and those with electrolytic sensors. In the paramagnetic type, oxygen passing through a fixed magnetic field (developed by permanent magnets in the detector head) interferes with the magnetic flux, thereby causing a small induced current. This current can then be related directly to the concentration of oxygen. This type of instrument is very sensitive to variations in pressure and temperature, but it offers sufficient accuracy for incident scenes. It is less often encountered than instruments with electrolytic sensors.

Electrolytic sensor-type analyzers have either an internal sensor (i.e., located inside the instrument casing), where the sample is drawn through the sensor by use of an aspirator bulb, or an external sensor that relies on diffusion of the air through the sensor head. In the latter type, the sensor head is located at the end of an electrical signal cord; in the former a hollow tube is used to draw remote samples to the sensor. Figure 6.2 illustrates a typical diffuser head.

Recent trends in this type of instrument have led to the development of small electronic units that allow diffusion of ambient air into the unit and its sensor. These instruments also allow for the facility of high- and low-level alarms and digital displays. Their principle of operation is similar to that described above.

As with all instruments, care should be taken to consult manufacturers' manuals prior to use. Particular attention should be paid to methods of calibration, limitations in use, and vapors that could cause erroneous readings or cell "poisoning."

3B. Combustible Gas Indicators

Combustible gas indicators (CGIs) measure the percentage LFL and give an indication of total combustibles present at an incident site. On most models there are two scales, 0–100 percent and 0–10 percent. There are many variations of this instrument, including some that give an LED/LCD readout. It is not within the scope of this book to discuss detailed operation of all the instruments. The "explosionmeter"-type instrument works on the same principles as the other, more advanced units, so it is used here for discussion purposes. The same precautions pertaining to this instrument apply to the other types.

A diagrammatic illustration of an explosionmeter appears in Figure 6.3.

The sensing element of the instrument is a metal filament heated by the electric current flowing through the unit (as supplied by six D-type cells). When mixtures of combustible gases in air are drawn over the filament, the gas oxidizes on it and causes the filament temperature to rise. This in turn causes a change in the balance of the instrument, which can then be measured.

Essentially the instrument is built around the principle of the Wheatstone bridge, the catalytic filament forming one of the resistances (R-1 in the diagram). The indicator is made ready for use by closing the switch and selecting "Voltage Adjust," the voltage adjust rheostat is then moved until the galvanometer needle is aligned with the voltage adjust mark. Next, the selector switch is turned to % L.F.L. and, while drawing fresh air through the instrument, the needle is adjusted to zero using the "zero adjust" resistor. The bridge is now in balance, the current from A to D via B being the same as from A to D via C, and the flow through the galvanometer from B to C is zero.

When hydrocarbon gases flow across the sensor filament, the temperature rises in proportion to the volume of gas, hence resistance also rises proportionally. This leads to current flow between C and D and a proportional deflection of the galvanometer needle, which can then be related directly to percentage LFL.

There are some precautions to be taken when using the instrument:

- Be sure that the flame arrestor gauzes are in position and are functional, otherwise the instrument may cause a "flashback." These gauzes have a tendency to clog over time and may have been erroneously removed.

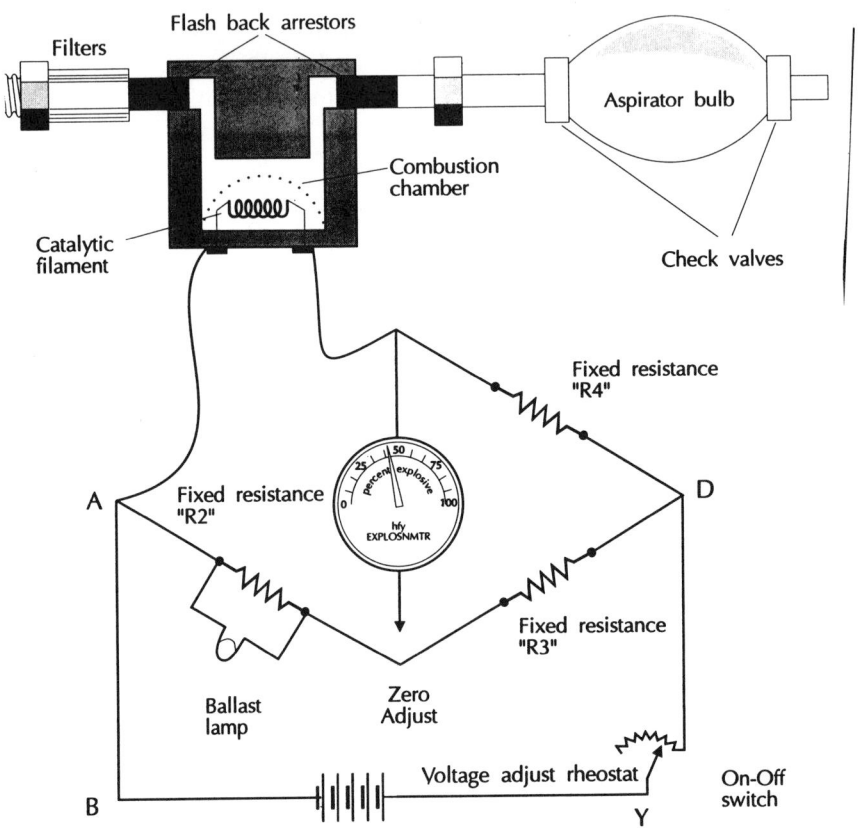

FIGURE 6.3. Combustible Gas Indicator

- Instruments are generally calibrated on hexane, pentane, or methane in air and hence give the truest readings using these gases. In oxygen concentrations below 11 percent, the instrument will not function; ideally it should be used only with 20.8 percent oxygen. An oxygen check should be carried out prior to using the CGI.

  If hydrocarbon vapors are of substances other than the calibration gas, corrections should be applied as shown in Table 6.1.
- Be sure that the instrument is regularly calibrated as per the manufacturer's manual.
- Constantly monitor the instrument during sampling. If an atmosphere that is too rich is sensed, the indicator needle will rapidly swing to +100 percent and then back to zero. If this is not noticed, an erroneous zero

TABLE 6.1 CGI Correction Factors

Combustible	Factor
Amyl chloride	1.25
Octanes	1.16
Xylol	1.12
Butyl acetate	1.11
Acetone	1.00
Butane	1.00
Hexane	1.00
Pentane	1.00
Toluene	1.00
Benzol	.91
Ethyl acetate	.91
Ethyl ether	.91
I.P.A.	.91
E.M.K.	.91
Ethyl alcohol	.77
Methane	.77
Ethane	.72
Methyl alcohol	.72
Propane	.72
C.O.	.57

will be obtained. Most instruments in use have a high-level alarm set at 20 percent LEL, which should negate this problem.
- In winter, beware of the problem of a high temperature differential. That is, if sampling a vessel, the temperature of which is 60°C (140°F) on a day when the ambient temperature is 0°C (32°F), as the hot vapor passes into the instrument, some of it will condense onto the instrument casing and flashback arrestors. This then means that the sample entering the filament chamber contains less hydrocarbons than the vessel and hence an erroneous reading will be obtained.
- If using an extension sampling hose, ensure that the instrument is purged with the sample (one aspiration per 2 ft of hose).
- Always purge the instrument with fresh air after use and be sure that it is switched off.
- To repeat, calibration is essential—a check using a gasoline-soaked rag is no substitute for a span check with calibrating gas.

3C. Colorimetric Detection Tubes

Many substances will not register on a CGI (e.g., sulfuric acid) or, even if combustible, cannot be detected at a safe level. For example, acrolein has an IDLH value of 5 ppm, which would not be detectable on a CGI.

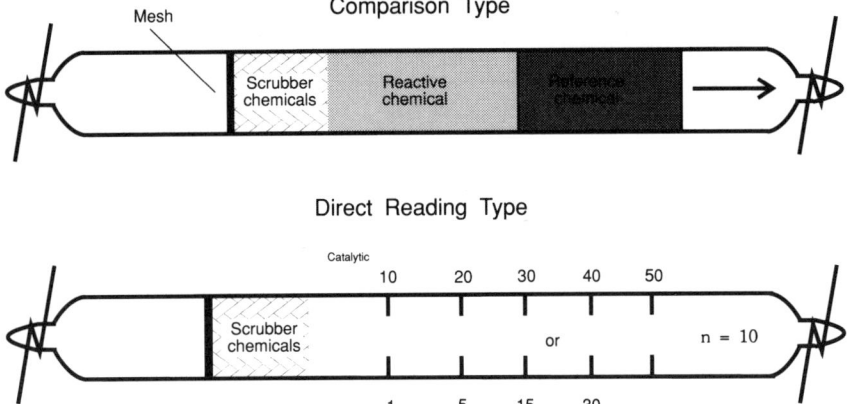

FIGURE 6.4. Colorimetric Sample Tubes

To measure very low concentrations and the presence of nonflammable vapors at an incident site, another system must be utilized. The two alternatives most often available are photo-ionization detectors and colorimetric detection tubes. The colorimetric tube system consists of a pump or bellows arrangement that draws a precise volume of air through a reactive chemical substance. Presence of the vapors being monitored for is indicated by staining of the reactive chemical. The degree of staining and the number of pump strokes can be related directly to the concentration in parts per million; exact correlation is determined after correction for factors such as temperature, pressure, and humidity. Two types of tube are in use. One relies on counting the number of pump strokes required to change the color of a reactive chemical to that of a reference chemical; most tube types encountered today rely on the substance being detected causing a directly readable change in chemical color. The scale of these may be linear or nonlinear, depending on whether or not the reaction is catalytic. These two types are illustrated in Figure 6.4.

The following precautions should be noted in the use of colorimetric tubes.

- All tubes have a limited shelf life; expiration dates are stamped on the box containing the tubes. Shelf lives of significantly less duration than stated will occur if the tubes are stored in a high-temperature location.
- Many tubes suffer interference from substances other than those being investigated. Incorrect staining may mean that the wrong material has been identified or that vapors of other materials are present. Consult the manufacturer's data sheets for further discussion of these effects.

- The reactive chemical in some tubes is toxic. Avoid inhalation of any exhaust gases and/or tube contents.
- The best accuracy of these systems is ±25 percent, and all values obtained must be treated accordingly.
- Ensure that the tubes in the box match the box label, especially if the seal on the box is broken.
- All tubes are sealed and the glass ends must be broken before use; caution in handling the tubes is necessary.

3D. Photo-Ionization Detectors

Recent trends in photo-ionization detectors (PIDs) have been the development of portable, hand-held units with sophisticated data logging and display capabilities. PIDs measure total photo-ionizable contaminants in a sample, but do not discriminate between substances. They will detect many organic and some inorganic materials. PID's are very sensitive and will indicate low concentrations of contaminants, in the ppm range. Typically, units have a minimum sensitivity of 0.5 ppm.

Photo-ionization detectors work on the principle of ionizing contaminants by exposure to ultraviolet light. A small charge is produced by this ionization that is proportional to the number of molecules ionized.

The energy required to cause ionization of a substance is measured in electron volts (eV). If the substance has a higher ionization potential than that of the ionizing lamp (typically 9.6 eV), it will not register. Manufacturers supply a list of substances that can be detected for a particular lamp. Further information on the ionization potential of a substance may be obtained from references such as the *NIOSH Pocket Guide to Chemical Hazards*.

In closing, a word of caution. The author has audited many spill response teams and industrial facilities. Many new instruments have detection capabilities offering three parameters, oxygen, combustibles, and "toxics." On several occasions instrument owners have stated that the "toxic" indication will detect the presence of any toxin. This is not the case: The "toxic" cell is substance specific for gases such as hydrogen sulfide, carbon monoxide, and carbon dioxide.

7
Tank Trucks

Many types of bulk liquid cargo tanks (tank trucks) are used in transportation. Because of their size and the quantity of materials handled, this chapter highlights their features and safety equipment, and offers some ideas of how to deal with these units in an emergency.

Tank trucks are made of a variety of materials, including mild steel, stainless steel, alloys, aluminum, and fiberglass. Depending on the materials handled, they may be lined, insulated, segmented, double-hulled, or pressurized.

Specifications for the construction of tank trucks are detailed in Title 49 CFR 173.33 and Section 393 of the Federal Motor Carrier Safety Regulations; these standards are also maintained in Canada.

1. SAFETY DESIGN FEATURES

Many features in the design and construction of tank trucks assist in the safe transportation. These include the following.

- Internal valves, designed to be housed within the tank shell so that even if external piping is broken the tank contents will not be lost.
- Shear sections on external piping so that, in the event of pipe damage, the piping will break at a predetermined point, where the least amount of damage can be caused. Such breakage occurs within 10 cm (4 in.) of the tank and ensures that the internal valve will remain intact.
- Rear-end protection, in the form of heavy-construction bumpers. This is designed to protect the tank shell and any piping at the back of the truck during a collision.
- Overturn protection. All upper tank openings are shouldered by welded

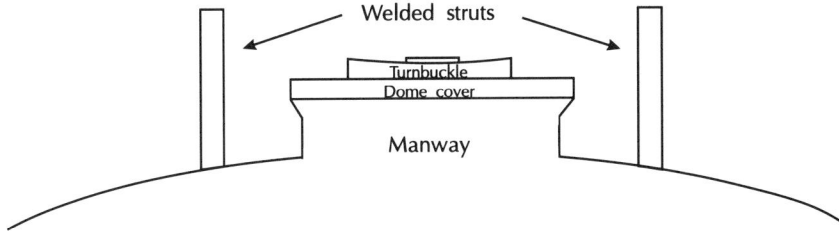

FIGURE 7.1. Overturn Protection

struts which are designed to protect the opening and its fittings, as shown in Figure 7.1.
- Overpressure/vacuum protection. All tank trucks, whether pressure or nonpressure, are fitted with safety venting and relief valves. Depending on the unit, this may take the form of spring-loaded or bursting-type units, or a combination of the two. However, where bursting-type units are fitted, they must be in series with a spring-type unit, not in parallel.
- Emergency shutdowns. All remote valves are protected by an appropriate shutdown mechanism. Such mechanisms are "fail safe" and normally work against a spring action. Remote valves are either mechanical, pneumatic, or hydraulic, and the type will dictate the exact method of operation. These are reviewed later in this chapter.

Emergency shutdown valves or levers are located at the transfer point and immediately behind the driver's position. In addition to manual operation, all three types are fitted with some sort of fusible (heat-induced) release system. Pneumatic and hydraulic systems are also fitted with frangible (break-off) releases. Most designs incorporate these two types in a fusible/frangible plug arrangement.

2. CARGO TANK TYPES

The type of cargo tank and the materials that it can carry are summarized in Table 7.1. Note that previous MC 300 series designations are being upgraded to DoT 400 series designations. This reflects changes in construction requirements, particularly with respect to safety relief and venting devices.

Figure 7.2 is generic in its approach and does not address specific requirements for the transportation of certain products. Such requirements appear within Title 49 CFR.

The following sections discuss some of the features found in each type of unit and the general layout.

TABLE 7.1 Summary of DOT/MC Specifications for Tank Trucks

Designation	Type	Materials Carried	Construction Materials
MC306/DOT 406	Non-pressure	Petroleum products	Aluminum
MC 307/DOT 407	Low-pressure	Chemicals	Aluminum Mild steel
MC 312/DOT 412	Low-pressure	Corrosives	Mild steel Stainless steel Lined
MC 331	Pressure	Gases, non-cryogenic	Aluminum Mild steel

2A. Specification 306/406 Tanks

Figure 7.2 shows the general arrangement of a 306/406 specifications trailer. These trailers are generally used for the transportation of petroleum products such as gasoline, diesel, and fuel oil. Depending on their required service, they may be fitted with transfer valving at the rear or midpoint. Such valving is normally designed to allow one coupling for loading and another for discharge. In multicompartment trailers there is a loading point and off-loading valve for every compartment.

Internal valve control is normally achieved by mechanical cable operation, with individual controls for each compartment. The emergency shutdown levers will trigger all valves for all compartments; they are located at the transfer coupling points and at the front left of the trailer.

These tank trucks are not designed to withstand pressure and are fitted

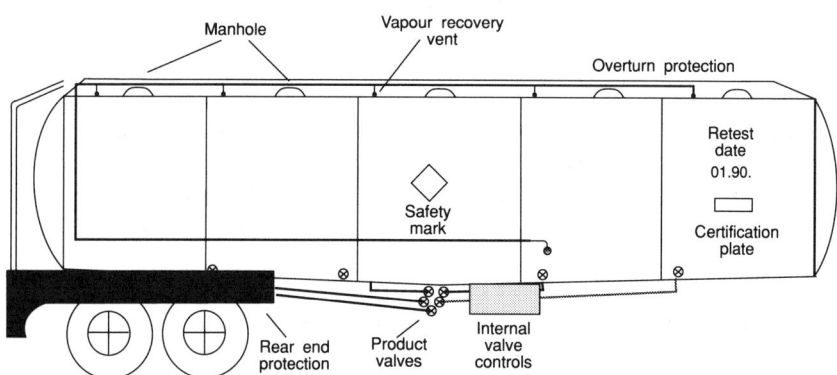

FIGURE 7.2. MC 306/406 Trailer

with safety relief valves designed to operate at 21 kPa (3 psi) pressure and 13.78 kPa (2 psi) vacuum.

In a rollover situation there are several techniques that may be utilized to remove product prior to uprighting the trailer. If the trailer is oriented in any plane other than upright, the bottom valving will be restricted in its capacity to drain the trailer.

When a trailer is rotated 90° it will only be possible to drain it to half-empty, assuming that the valving is reachable. Once the trailer is half-empty, the manholes will need to be opened and a pump suction introduced into the liquid. This is illustrated in Figure 7.3.

Great care must be taken when utilizing this technique. When the dome lid is first opened, a considerable amount of product will run into the portable tank until the liquid is below the dome lip level. Drainage of this liquid should be done by opening the dome on the turnbuckle arrangement slowly and controlling the flow of product into the portable tank.

Obviously, where volatile components (e.g., gasoline) are being handled, all due safety precautions against ignition contact and health precautions against vapor and liquid exposure should be utilized. Remember, once the dome lid is opened, the vapor space will probably be within the flammable range.

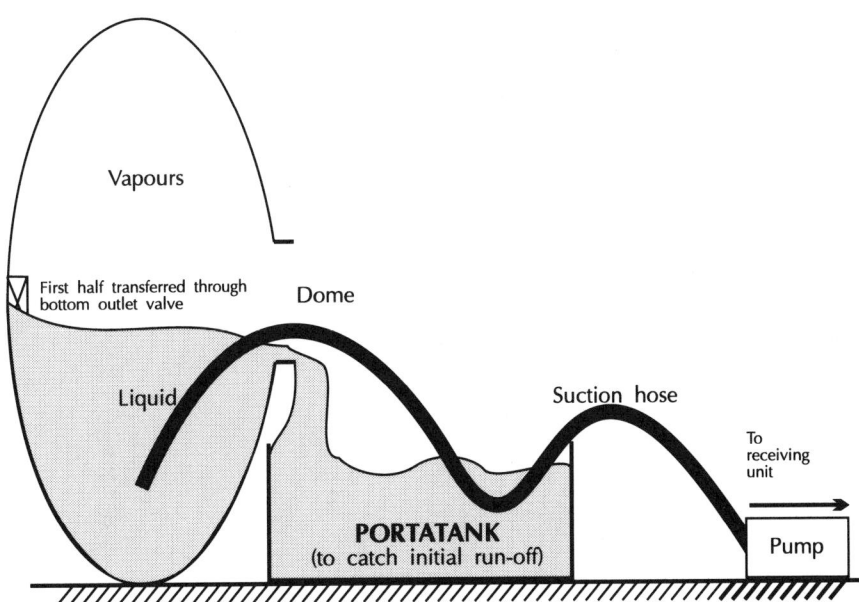

FIGURE 7.3. Transfer at 90° Rotation

162   Tank Trucks

One likely ignition source when using this technique is a static spark. Ensure that both units, the portatank pump, and lines are all bonded one to the other and that the whole arrangement is grounded.

In situations where the conventional valving is not accessible a patent device will need to be used to allow access to the tank. There are several manufacturers of dome cover devices such as that illustrated in Figure 7.4.

When a trailer is completely overturned, there are two options open. One is to remove the internal valve, not an easily achieved objective. Alternatively, with aluminum-shelled trailers, it is acceptable to drill a 2-in. (5-cm) hole in the shell and feed a hose suction in through the hole.

Figure 7.5 shows a typical layout for gasoline transfer of a rolled-over unit.

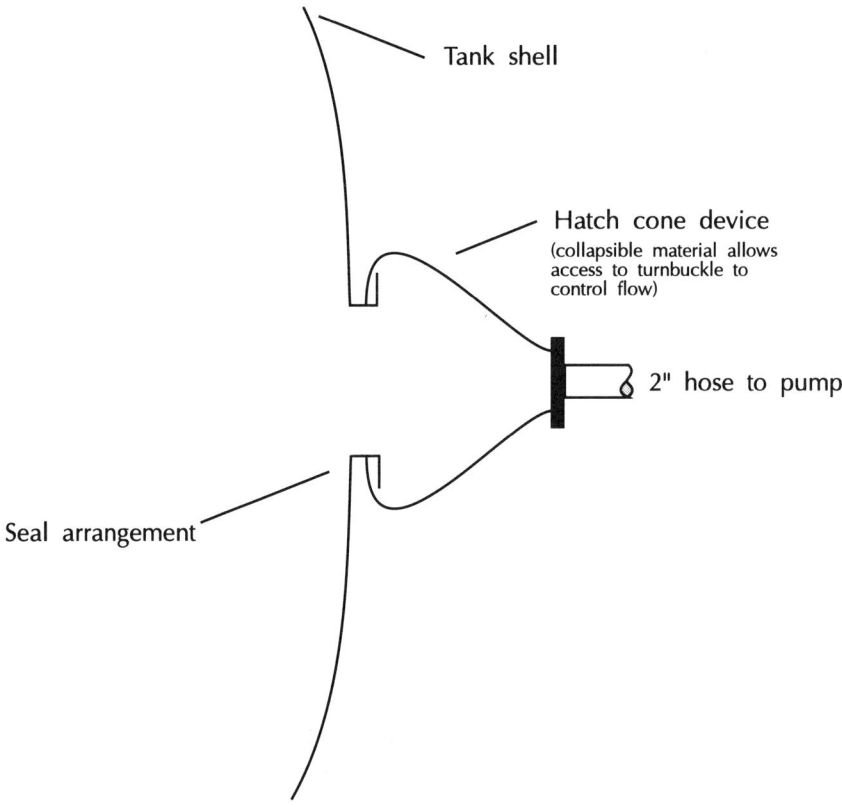

FIGURE 7.4.   Hatch Cone Device

2. Cargo Tank Types   163

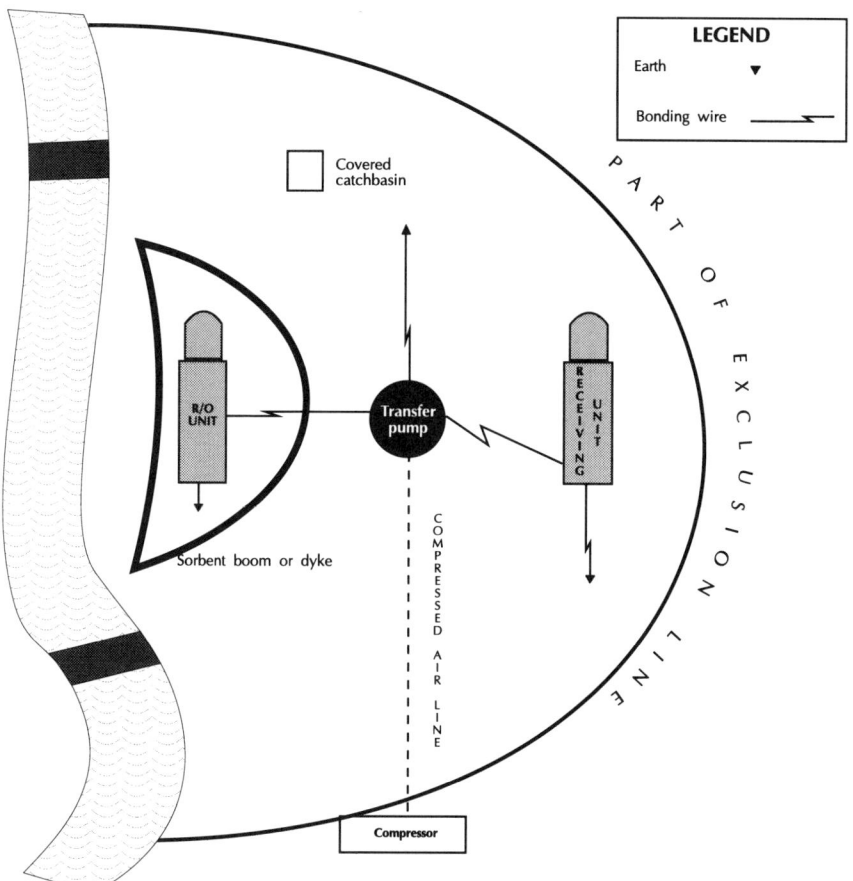

FIGURE 7.5.   Flammable Liquid Tank Truck Rollover: Transfer Setup

2B. Specification 307/407 Trailers

Figure 7.6 shows the general arrangement for a 307/407 specification trailer. These trailers are designed for the transportation of chemicals. Round in cross section, they may be constructed of aluminum, mild steel, or more often stainless steel. Strengthening is normally achieved by the use of external support and stiffeners, to facilitate cleaning.

These trailers may be insulated or uninsulated. If insulated, they are usually sheathed and have a characteristic horseshoe section when viewed from the rear.

Specification 307 trailers are usually designed to haul one product and

164   Tank Trucks

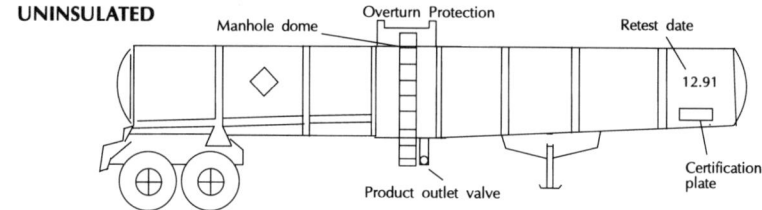

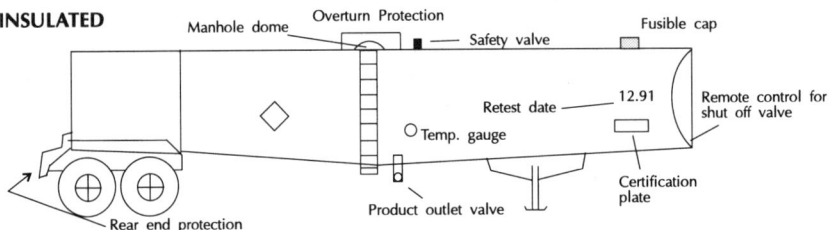

FIGURE 7.6.   MC 307/407 Specification Trailer

are fitted with loading and unloading arrangements either at the back or in the middle. The slope of floor on the underside is adapted to facilitate this.

The minimum working pressure for a specification 307 trailer is 172.25 kPa (25 psi). When the trailer is required to work at pressures above 344.5 kPa (50 psi), the tank is constructed to ASME code requirements.

Valves on these trailers are typically hydraulically operated, but may be mechanical or, in some units, pneumatic.

In rollover situations, the same considerations to product removal apply as to specification 306 trailers.

2C. Specification 312/412 Trailers

Figure 7.7 shows the general arrangement for 312/412 specification trailers. These trailers are designed for the transportation of acid/caustic solutions. Loading and unloading is achieved by either top or bottom access. For top unloading, by pressure, an extension tube extends to the tank bottom, as shown in Figure 7.8.

Bottom outlet valves may also be fitted, which will either have an internal shutoff valve (as with 306 specification trailers) or, if the product to be carried may cause sedimentation around the valve, an external valve with horizontal guards surrounding the piping.

2. Cargo Tank Types   165

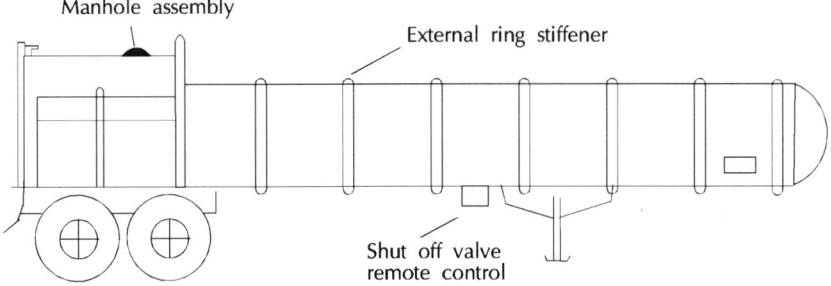

FIGURE 7.7. MC 312 Specification Trailer

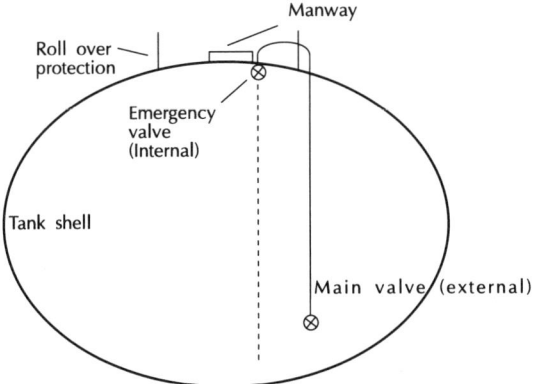

FIGURE 7.8. Dip Tube Arrangement

In rollover situations the considerations for product removal are the same as before, the presence of the extension tube making no difference in terms of how to remove product.

2D. Specification 331 Trailers

Figure 7.9 shows the general arrangement of a specification 331 tank. These tanks are designed to carry gases under pressure. Typical gases carried in a standard specification 331 unit include propane, butane, and ammonia. There are specific requirements for chlorine, carbon dioxide, and several other gases relating to construction and closure devices, which are referenced in 49 CFR. All specification 331 units have a minimum design pressure of 689 kPa (100 psi) and a maximum of 3445 kPa (500 psi).

The lines and valving on 331 trailers consists of liquid, vapor, and spray

166 Tank Trucks

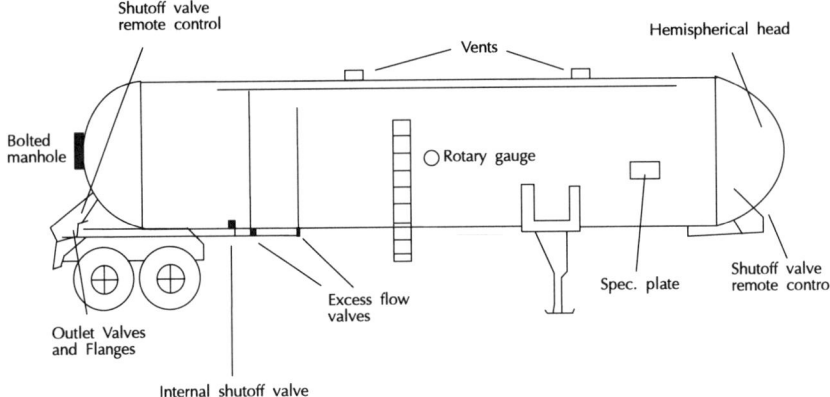

FIGURE 7.9. MC 331 Specification Trailer

fill. As shown in Figure 7.10, each of these three lines has a different arrangement inside the tank. The liquid line connects with an internal valve and feeds directly to the tank bottom, similar to a specification 306 arrangement. The vapor line is a "standpipe" arrangement that communicates with the upper areas of the tank space. The spray fill line connects to a slotted bar that runs the length of the tank; this facility is used during initial cooldown of a trailer.

In rollover situations consideration to using the various arrangements should be made; i.e., if the unit is rotated 180°, discharge may be achievable by the spray bar, provided the nonreturn valve can be secured.

2E. Nonspecification Trailers

In addition to specified trailers, several nonspecification trailers are used to transport dangerous goods. An example of this is an asphalt trailer.

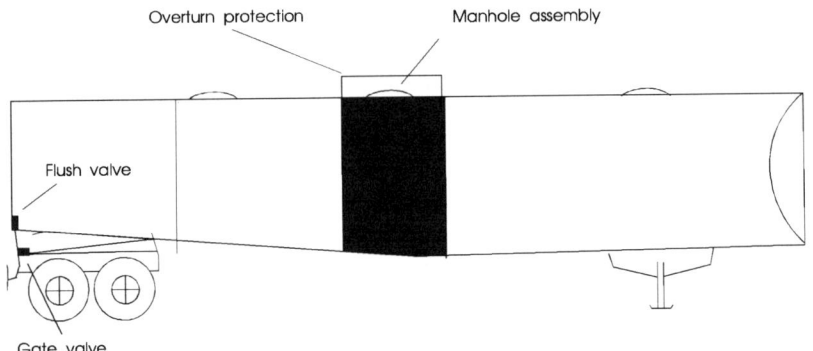

FIGURE 7.10. Asphalt Trailer

Figure 7.10 shows the general arrangement of an asphalt trailer. These trailers are constructed of aluminum or, more typically, mild steel. The trailers are heavily insulated so that they can transport liquid asphalt, typically around 150°C (300°F), without the need for steam heating. Some trailers are fitted with heating coils, but these are specially adapted and require an external steam source.

Unlike specification trailers, there are no particular venting or emergency shutoff arrangements for asphalt trailers. Indeed, unloading is normally achieved through a simple gate valve at the trailer rear. However, when low-flash-point asphalt is carried, the trailer must be fitted with 306/406 specification relief valve and emergency control fittings.

In rollover situations, the prime concern is cooling and consequent solidification of the product. Obviously, escaped product can be cooled with water, thus creating an effective "dam" to further product movement. Transfer to receiving units is difficult and needs specialized gear wheel-type pumps.

On nonspecification trailers, vacuum breakers may not be fitted, so it may be necessary to drill through the tank shell prior to pump-out. One advantage of this trailer type is that suction hoses can be fed directly through the bottom outlet valve without disassembly.

3. VALVES

Many different types of valving arrangements are found in tank truck trailers. This section discusses some of them.

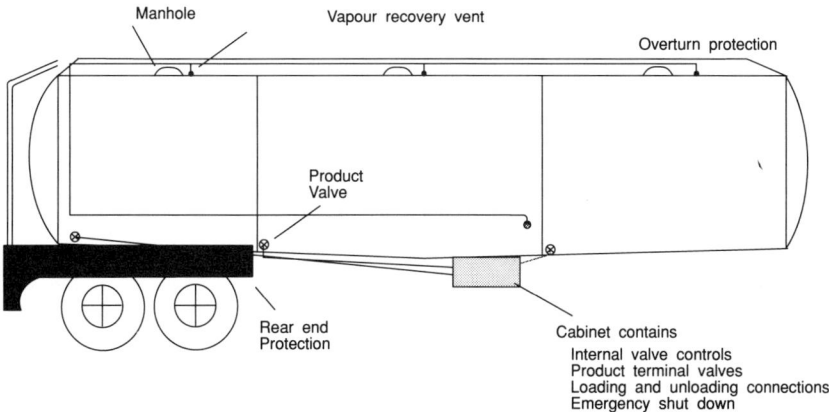

FIGURE 7.11. MC 306 3 Compartments: Typical Valving Layout

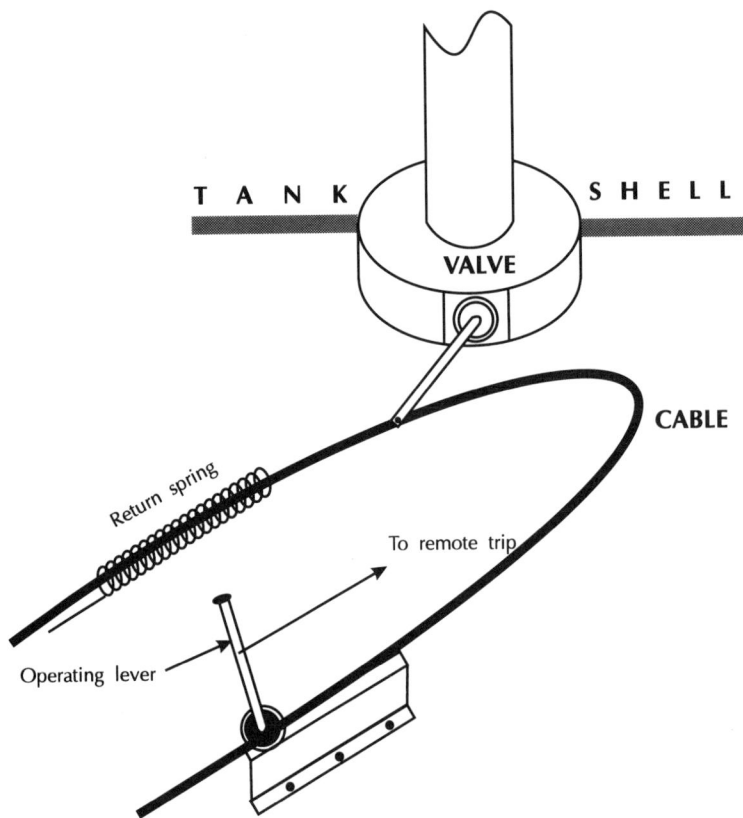

FIGURE 7.12. Cable-Operated Valve

3A. Specification 306/406 Tank Valves and Fittings

Specification 306/406 units are generally fitted with cable-operated internal valves and hand-operated gate valves at the coupling point. Figure 7.11 illustrates the location of the various valve types, and Figure 7.12 depicts the arrangement of an internal valve.

The remote lever gate valves and loading connections may be at the midpoint or after end of a trailer. In many units the valves are housed in a "cabinet"-type enclosure to protect them from damage during routine operations. Some arrangements are such that closing the cabinet door automatically trips all the cable-operated valves.

The terminal valves on these specification units are normally of the "Y"-arrangement type. This is illustrated in Figure 7.13. The valve allows for both loading and unloading through the bottom of the unit. Note that

3. Valves   169

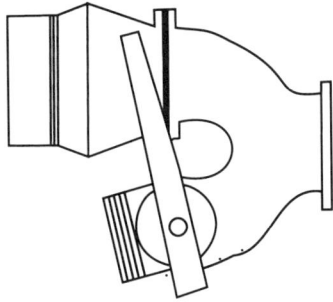

FIGURE 7.13.   "Y" Terminal Valve

the valve is spring loaded in the closed position and must be set manually before attempting to transfer product. Obviously, the internal cable-operated valve must be set likewise.

3B. Specification 307/407 Trailer Valves and Fittings

As discussed previously, specification 307/407 units are normally designed to transport a single product, with valves located either at the back or middle of the trailer. The valving system used is generally a hydraulic, occasionally pneumatic, internal valve with a hydraulic valve fitted at the terminal point. Figure 7.14 shows the general arrangement for an internal

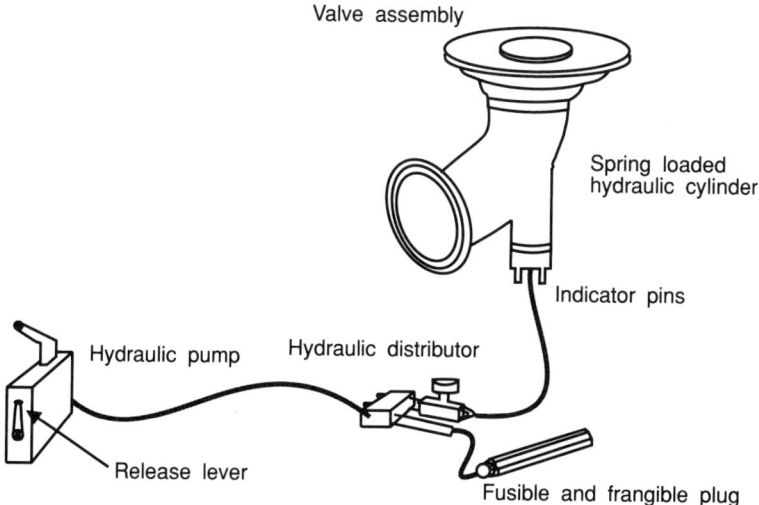

FIGURE 7.14.   Hydraulic Valve Arrangement

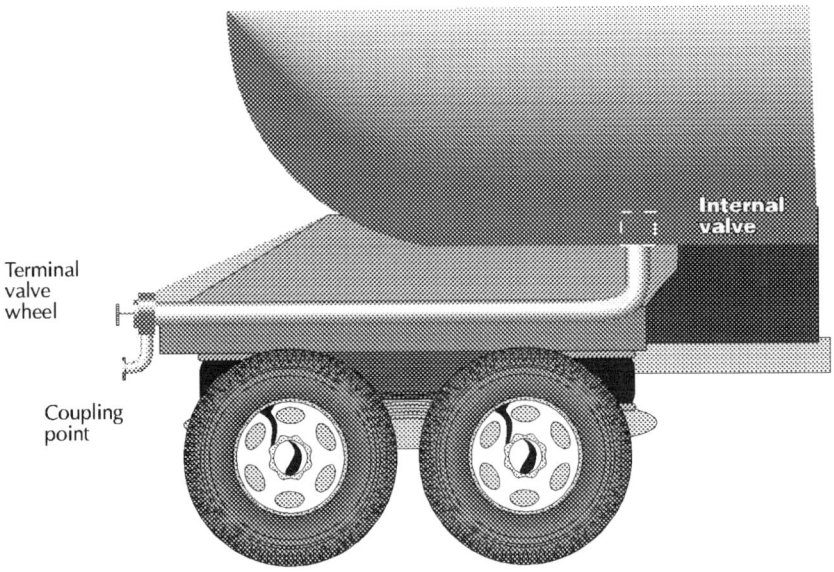

FIGURE 7.15. Terminal Valve Arrangement

hydraulically operated valve. The valve is opened by moving the release lever to the "closed" position and cranking the hydraulic pump. This in turn causes the valve to open. Opening the release lever causes depressurization of the line, and the spring action on the valve causes it to close.

The hydraulic line is paralleled by a fusible/frangible plug (i.e., one that will melt or can be snapped off).

Figure 7.15 shows the arrangement of the terminal valve. This valve, commonly known as a "hydrolet" valve, is designed primarily with ease of overhaul and cleaning in mind. The valve is constructed of stainless steel with a Teflon packing.

8
Rail Cars

Joe P. Riddle

Railroad tank cars have been found to be one of the easiest and least costly methods of shipping liquid products throughout the world and particularly in North America. Nearly 200,000 tank cars currently in service carry a variety of materials from corn syrup to highly toxic and explosive chemicals. Responders to incidents involving tank cars must presume the presence of hazardous materials, although many tank cars contain products used in everyday consumables and pose no threat to responders.

1. GENERAL DESCRIPTION

A tank car is nothing more than a cylindrical storage tank sitting on a set of wheels which allows it to be pulled on a pair of rails. All cars have common terms that are used to describe their features. Figure 8.1 shows a hypothetical car in order to illustrate the common features and terms. The framework which helps support the tank may take a variety of forms but typically involves the use of "sills" which provide support beneath the tank and "bolsters" which provide strength in the area of the tank which is carried by the wheels of the car, referred to as "trucks" (Figure 8.2). The attachment of the tank to the trucks is accomplished by a steel "center pin," which basically means that the tank structure is able to fall free or be lifted off of the trucks. The sight of the trucks being separated from the car itself may appear unnerving at a derailment, but the unit is designed in this manner. "Couplers" are knuckle-shaped devices which allow railroad cars to be connected to one another. The union of couplers between cars allows the couplers to slide vertically as cars are moved,

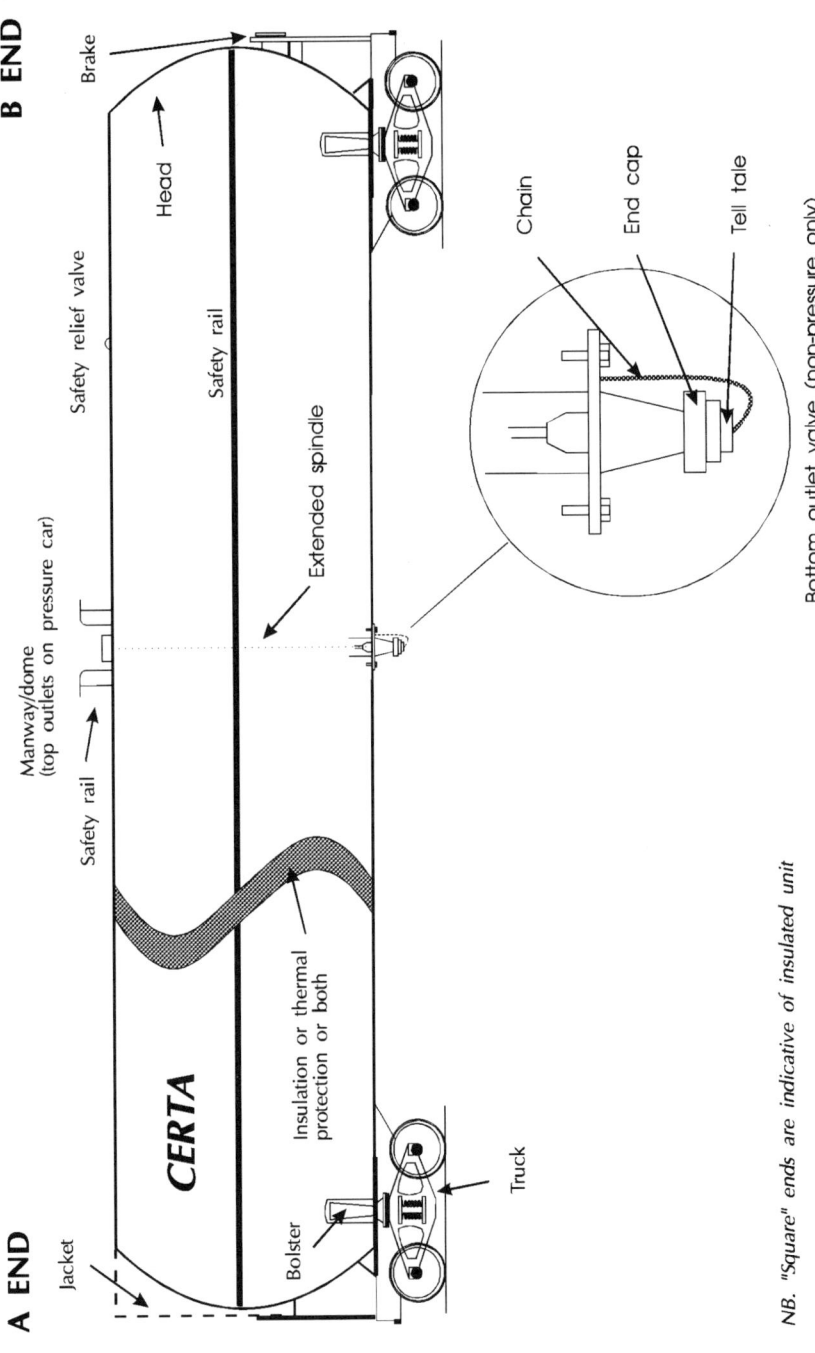

FIGURE 8.1. Tank Car Nomenclature and General Layout

1. General Description   173

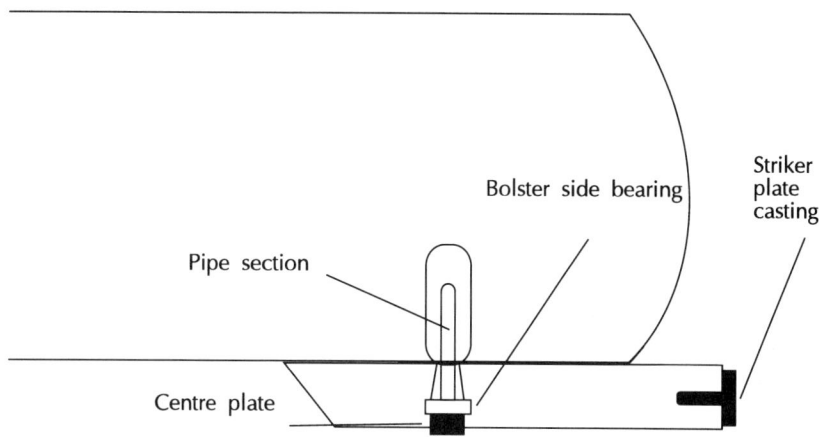

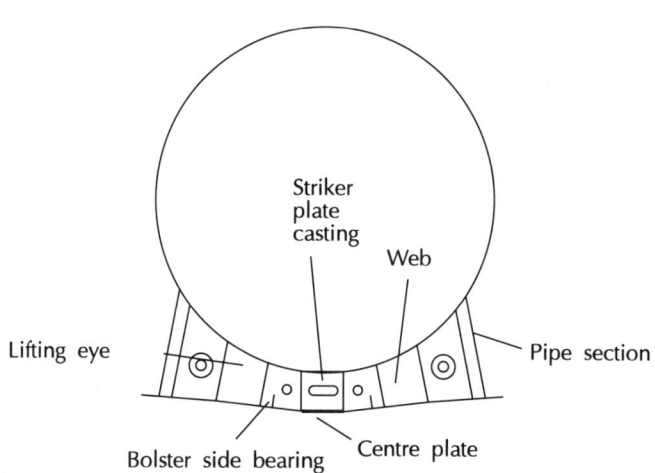

FIGURE 8.2.   Bolster and Support Arrangement

thus preventing binding. However, a series of accidents in the 1970s was caused by a vertical separation of the couplers and punctures to the "heads" of tank cars from adjacent cars' couplers. Retrofitting of tank cars carrying hazardous cargos has resulted in the use of shelf couplers, which prevent the vertical separation, and reinforcement of the heads of the tank cars.

2. RESPONSE TO TANK CAR INCIDENTS

As in all hazardous materials incidents, a top priority of responders is identification of the substances involved. In railroad incidents, this is best accomplished through the use of "waybills" or the "manifest." Each railroad car is identified by the use of the car's reporting marks, which consist of several letters followed by several numbers, as shown in Figure 8.3. A waybill is issued for each car, similar to a bill of lading for trucks, indicating the shipper, consignee, contents, and other pertinent information on the specific car. A manifest is a list of the cars in a train, in the order in which they are arranged in the train. The waybills and manifests will be in the hands of the conductor on the train, who can assist in the identification of the cars. Should an incident involve tank cars not in a train, a yardmaster or other railroad official will have the waybills.

Tank cars carrying hazardous materials must comply with placarding regulations and should be placarded on both sides and ends of the car. One variation of the placarding requirements involves the use of the "residue" placard. Once a tank car has been drained of product, the placards are typically flipped over and the bottom portion of the placard will be colored black with the word "residue" written in white. This is to inform emergency responders of the fact that the tank is not full, but still poses a threat to responders.

Some hazardous commodities carried in tank cars are required to have the name of the chemical stenciled on the side of the car in 10-cm (4-in.)-high letters. This allows responders at least to partially identify the materials from a safe distance.

If unable to identify a car's commodity by these means, responders can contact CANUTEC or CHEMTREC. Supplying one of these agencies with the car's reporting marks may permit them to track down the car's owner (usually not the railroad) and ultimately determine the car's contents.

Response to tank car incidents poses some unique challenges to the responder. If not already aware of the incident, the railroad whose track the car is on must be notified immediately in order to prevent car movement and to make notification to shippers and authorities. Of particular concern to responders should be the possibility of multiple chemicals being involved, particularly in a derailment situation. Hazardous materials cars are frequently grouped together in trains, and a derailment could result in damage to several cars and losses of different chemicals. An unstable situation such as this must be approached cautiously, being aware of the possibility of violent reactions between products. Another concern involves the damage to the tanks caused by the accident. Dents, scores,

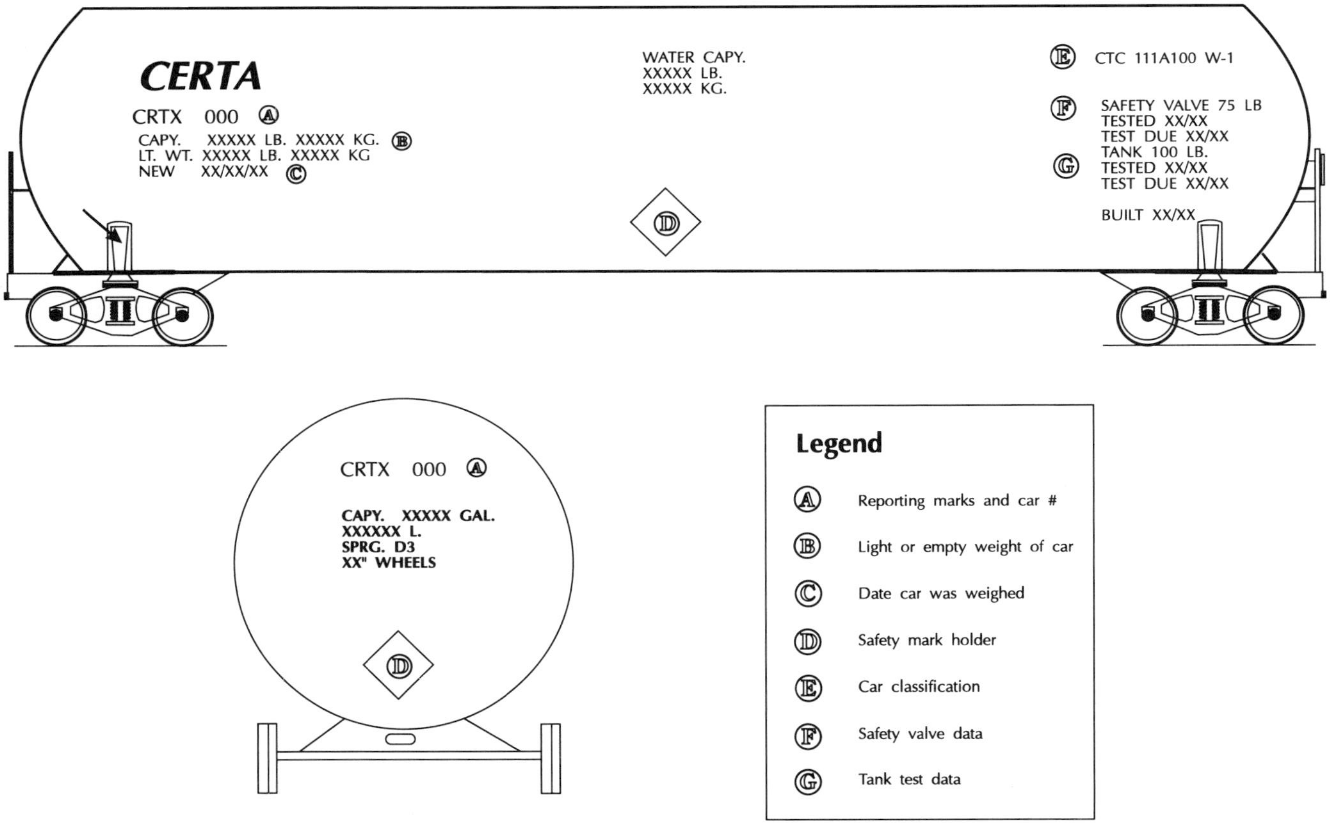

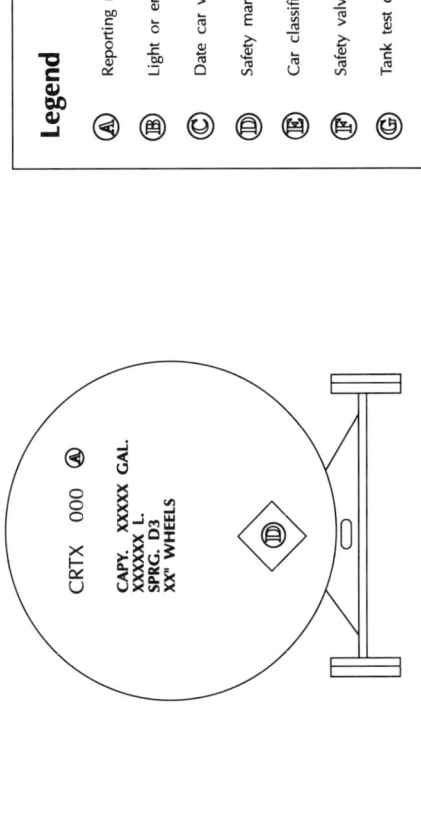

FIGURE 8.3. Tank Car Markings

and gouges to tanks, particularly those carrying compressed gases, can result in an unanticipated and unexpected total failure of the tank. BLEVEs are a very real hazard in multiple tank car accidents where there is fire.

3. TYPES OF TANK CARS

Tank cars are built under specifications determined by the Canadian Transport Commission's Regulations for the Transportation of Dangerous Commodities by Rail, the Association of American Railroad's (AAR) Specifications of Tank Cars, and the U.S. Department of Transportation's Title 49 CFR. The two fundamental divisions of tank cars are pressure and nonpressure specifications. Both types have several inherent features which will aid responders.

3A. Pressure Tank Cars

Pressure tank cars are used for the transport of flammable and nonflammable gases as well as Class A poisons. Liquid products which possess significant hazardous properties, such as ethylene oxide, sodium metal, and hydrofluoric acid, may also be carried in pressure cars. There are several specification categories for pressure tank cars, although the most common are the DoT-105 and DoT-112 styles.

Pressure tank cars nearly always have all the fittings and appurtenances located on top of the tank within a manway protective housing. This housing offers limited protection to the fittings in a derailment. The tank will have convex heads to hold the pressure, which can range from 689 to 4134 kPa (100 to 600 psi) depending on the commodity. Pressure tank cars hold from 22,740 to 159,180 liters (6000 to 42,000 gallons).

Valves and fittings which may be found on pressure tank cars include the following.

- Liquid valves. Most pressure cars have two liquid valves of 2 in. in size. These are attached to liquid eduction tubes that run to the bottom of the tank for loading and removing liquid. These eduction tubes will likely have excess flow check valves to automatically shut off the flow of product should a hose rupture during transfer.
- Vapor valves. Generally, one vapor valve will allow the return of vapors to the tank during transfers. A notable exception to this is chlorine tank cars, which are DoT 105 standard. These cars utilize 2.5-cm (1-in.) valves and have two vapor valves instead of one.

- Safety valves. Spring-loaded pressure relief safety valves are set to relieve at 75 percent of a tank's test pressure.
- Gauging devices. There are several styles of devices to measure the liquid level in a pressure tank car. Some of these involve opening a small valve and sliding a tube up and down seeking the level of liquid. Others utilize a magnetic float device.
- Sample Line. A 0.625-cm (0.25-in.) valve attached to a liquid eduction tube is sometimes present for taking a sample of the commodity from the tank prior to unloading.
- Thermometer well. A closed pipe leading into the liquid space with a cap is used for measuring the temperature of the commodity. A thermometer is lowered into the pipe to take the reading.

3B. Nonpressure Tank Cars

Nonpressure tank cars are used to carry liquids including flammables, corrosives, and poisons. Nonpressure cars are also used in the shipment of many products that are not hazardous, so responders must be careful not to jump to conclusions when faced with a tank car incident. Nonpressure cars are typically of thinner construction than pressure cars, ranging in test pressures from 241.15 to 689 kPa (35 to 100 psi). Capacities range from 15,160 to 170,550 liters (4000 to 45,000 gallons). Categories of nonpressure tanks are numerous, but the most common styles are DoT-103 and DoT-111. DoT-103 styles are typified by a large expansion dome on top of the tank to allow for liquid expansion while shipping. DoT-111 and pressure tank cars are filled only partially, allowing for expansion into the vapour space. DoT-111 cars are the most common style of tank car in service today. Nonpressure tank cars may be insulated and can be equipped with heating coils, interior or exterior, with steam and condensate fittings at the bottom of the tank.

Many of the valves and fittings found on pressure tank cars are also present on nonpressure tanks. However, nonpressure tank fittings generally are not housed under one protective cover. Fittings which can appear on a nonpressure car that are not found on a pressure tank include the following:

- Bottom outlet valve. A bottom outlet valve is used for draining a tank when unloading. It may be controlled by a valve handle beneath the tank or by a top-operated handle which travels from the top of the tank through the product to the workings of the valve.
- Manway. A manway that is hinged and with a bolted cover can be opened for tank cleaning and for loading the tank car.

- Safety vent. Some nonpressure tanks (particularly those carrying corrosives) will be equipped with a safety vent instead of a safety relief valve. The vent assembly is equipped with a frangible disk designed to fail at a set pressure. This disk does not reclose after it has failed, as a relief valve does.
- Vacuum relief valve. This valve opens to allow air into the tank while unloading the tank through the bottom outlet or through a liquid valve to prevent the creation of a vacuum within the tank and possible collapse of the vessel.

9
Intermodal Tank Containers

Joe P. Riddle

Intermodal tank containers constitute a new generation of bulk liquid container which is experiencing significant growth in North America. Tank containers, as they are most often called, have been in use in Europe since the 1960s and have increased in use throughout the world since their inception. The concept of a tank container is very similar to that of other intermodal containers: the ability to be loaded once at a shipper's facility and then be transported in a single vessel by way of highway, rail, and water carriers. This produces a much more economical and safer form of transport, as products do not have to be transferred from vessel to vessel as modes of transport change. Tank containers have found their biggest use in overseas shipments, and although they are able to carry almost any commodity, they frequently carry the more exotic, specialized products from different parts of the world. A response to a hazardous materials incident involving an intermodal tank container will present some unique challenges to the responder.

1. DESCRIPTION OF CONTAINERS

A tank container typically consists of a single, cylindrical-shaped vessel that is noncompartmentalized. The tank, or barrel, is mounted inside a supporting frame which allows the containers to be stacked aboard container ships and to be attached to each other and to truck beds and flatcars by the use of corner castings. The frame allows for the lifting and movement of the vessel by appropriate cranes and forklift devices. The frame also includes walkways for work atop the vessel and a built-in ladder to access the top. The framework is generally of one of two styles. The box-type frame encloses the tank in a cagelike framework and the tank itself is not an integral part of the frame. In the beam-type frame, frame structures are attached to the tank, making the vessel itself part of the framework. Frames are made of steel and generally are 20 ft long, 8 ft wide, and 8 1/2 ft tall, although there are no requirements regarding the size of the tanks. They may be of many different dimensions.

The tank itself is almost always made of stainless steel, although some tanks may be made of mild steel, aluminum, or other alloys. Although standards allow latitude in size, most intermodal tanks are about 24,000 liters (6340 gallons) in capacity.

Standards for intermodal tank containers involve many organizations, since the containers are used internationally. The International Maritime Organization (IMO) issues the standards for ocean transport of dangerous goods in tank containers in the International Maritime Dangerous Goods Code (IMDG code). This code also covers the design, testing, and construction of intermodal tanks and is widely accepted by most countries that import or export tank containers. The Canadian Transport Commission, the U.S. Department of Transportation, and other regulatory bodies have adopted standards for the handling, labeling, and shipping of tank containers in their respective jurisdictions.

IMO standards for tank containers divide containers into various types. The most common types of tank containers which responders are likely to face are the IMO type 1, IMO type 2, and IMO Type 5 tanks.

- An IMO type 1 tank container is designed for maximum allowable working pressures (MAWP) of between 175 kPa (25.4 psi) and 680 kPa (100 psi). They may transport both nonhazardous and hazardous materials including poisons, corrosives, and flammables. An IMO type 1 is the tank container most likely to be encountered by responders.
- An IMO Type 2 tank container is designed for MAWPs of between 100 kPa (14.5 psi) and 175 kPa (25.4 psi). Although some hazardous substances may be transported in type 2 containers, such as pesticides, solvents, and mild corrosives, they predominantly carry nonregulated materials such as food additives.
- An IMO Type 5 tank container is designed for MAWPs of over 680 kPa (100 psi) but not to exceed 340 kPa (500 psi). These tanks are used for transporting liquefied gases under pressure and also some of the more exotic liquids.

2. TANK CONTAINER FITTINGS

Since intermodal tank containers are international in use and manufactured in several different countries, not all of the following fittings will exist on

2. Tank Container Fittings   181

every tank container encountered. However, responders should be familiar with these, as they are likely to be present on a tank container involved in an incident. Since many tank containers are manufactured in Europe, it is quite possible that all fittings and outlets will be metric in size. Unusual thread designs may be encountered.

- Bottom outlet valves. Two externally operated, bottom outlet valves are required when a tank container is carrying hazardous materials. These are normally butterfly-type valves. The first is an internal valve, also known as a foot valve, which is connected in series with an external valve. The internal valve is equipped with a remote shutoff for use in an emergency. This emergency shutoff is located on the right-hand side of the tank as one is facing the discharge end of the tank. The external valve is equipped with a liquid-type closure such as a blind flange, screw cap, or possibly a cam-lock cap. The outlet size is normally about 7.5 cm (3 in.). Regulations prohibit some tanks carrying hazardous materials from having bottom outlet valves.
- Manhole. A 45.72- to 61-cm (18- to 24-in.) manhole is typically located on the top of the tank near the center. It has a hinged cover secured by wing nuts with a replaceable gasket. A dipstick may be in place inside the manhole to measure the liquid level of the tank. Measurements from the dipstick are compared to a tank calibration chart to estimate the volume of product in the tank. The manhole as well as all top fittings are surrounded by a spillbox which will drain spilled product as well as rainwater through one or more drain pipes.
- Top loading valves. A top loading valve is typically attached to a removable eduction pipe which runs down near the bottom of the tank. This valve may be a ball type or a butterfly type. Some tank containers do not have an eduction pipe and valve and are equipped only with a blind flange at this fitting. This fitting is approximately 7.5 cm (3 in.) in size.
- Air line connection. An air line connection is found near the top loading valve for the purpose of pressure unloading, blanketing the contents with an inert gas, or for vapor return. This connection is about 2.5 cm (1 in.) in size.
- Pressure/vacuum relief valves. Two combination pressure/vacuum relief valves protect the tank from pressure damage. Due to the thin construction of the tank (5–10 mm), as little as 5.17 kPa (0.75 psi) negative pressure operates the vacuum relief. The pressure relief on IMO type 101 and 102 tanks will operate from 175 to 600 kPa (25.38 to 87 psi) depending on the tank type and working pressure. This valve may be protected with a rupture disk between the tank and the valve to

prevent the commodity from corroding or plugging the valve. If the tank has a rupture disk, a pressure gauge will be present to determine if the disk has failed.
- Thermometer. A built-in thermometer is usually provided for measuring the temperature of the commodity.
- Insulation. IMO type 1 and type 2 tank containers are equipped with heating devices to keep the commodity under temperature control. Steam or another hot fluid may be circulated through piping located between the tank and external jacket. Tanks equipped with steam heating have inlet and outlet connections located at the end of the tank near the bottom outlet valves. Some tanks utilize electric heat by way of coils located between the tank and outer jacket. Voltages of 220 or 440 V are used to keep the product warm.
- Document tube. A tube is located on the side of the tank for placement of necessary papers such as cleaning certificates.

3. TANK CONTAINER MARKINGS

Tank containers may have a large number and variety of markings on the vessel. Although all of these have some significance, responders should be most familiar with the tank identification number, the dataplate, and the hazardous materials placards.

- Tank identification number. Each intermodal tank container is equipped with reporting initials and numbers. The initials indicate the ownership of the tank and the number identifies the specific tank. These initials and numbers are found on each side and end of the tank. These identification marks can be used by responders to determine the tank's contents using shipping papers or computer databanks such as CANUTEC, CHEMTREC, or through the carrier of the tank container.
- Dataplate. Tank containers have permanent dataplates attached to the tank or frame. Information on the dataplate includes tank type, tank capacity, tank pressure, materials of construction, and connections dimensions. The dataplate is usually found near the end of the tank with the bottom outlet valves.
- Hazardous materials placards. Tank containers carrying hazardous materials must be placarded according to the regulations of the country they are in, including the U.N. identification number. Obtaining these numbers may be difficult because of confusion with the other markings present on the container. It is also important to remember that different countries placard commodities differently, so a knowledge of other placarding systems is helpful.

4. RESPONSES TO INTERMODAL TANK CONTAINER INCIDENTS

Hazardous material releases from tank containers typically are not different than those involving other bulk containers. Identification of the spilled materials remains a high priority and can be accomplished through the use of shipping papers, placards, and by interviewing individuals at the scene. Work zones must be established and proper protective clothing selected and worn based on the nature of the materials and the size of the release.

One specific safety aspect involving tank containers involves accessing the top of the tank if necessary. The ladder built into the framework of the vessel does not lend itself to being climbed easily by personnel wearing personal protective equipment, particularly Level A equipment. Response personnel who need to access the top of a tank container should utilize an extension ladder or other safe means of reaching the top. Also, the working platforms on top of container tanks often utilize sharp gratings, which can tear protective clothing. There are no handrails to prevent personnel from accidentally stepping off the sides or ends of the tanks.

Releases or leaks from tank containers nearly always involve the fittings. Responders should remember that fittings may be metric sizes and carry appropriate tools. Some of the more common problems that responders might encounter include the following.

- Manhole gasket. The manhole gasket is the most likely leak point on a tank container because of the quality of the seal made with the manhole cover. Vapor leaks from manholes can be corrected by tightening the manhole wing nuts or replacing the Teflon-impregnated asbestos gasket with a new gasket. A liquid leak from an upright container at the manhole is indicative of an overfilled container.
- Top unloading valve, air line connection. Leaks from valves may indicate a poorly seated valve, debris in the valve, or a poorly connected valve or blind flange. Any liquid leakage at the top of the container may indicate an overfilled container. Responders should note that liquid leakage at the top of the tank will be caught in the spillbox and drain through the drain tubes. These drain tubes run between the tank and insulation jacket on insulated tanks and may give the appearance of a leak from inside the jacket.
- Bottom outlet valves. Leakage from bottom outlet valves is generally indicative of a poor seal of the internal or foot valve. This spring-loaded valve may be opened and allowed to slam shut to attempt to improve the seal. The remote shutoff handle can also be pulled to perhaps seat the valve. The internal valve cannot be opened from the remote shutoff.

Responders should always consider pulling the remote shutoff handle in a large spill from the bottom outlet to quickly secure the spillage with a minimum of risk to responders.
- Relief valves. Intermittent operation of relief valve in a nonfire situation is an almost sure sign of an overfilled vessel. Continued operation of a relief valve may indicate a failure of the relief valve. Some relief valve assemblies will leak if the rupture disk beneath the valve has failed. The pressure gauge can be checked to confirm this.
- Heating systems. Leaks from the heating coils of a vessel can be confused for a product leak. Steam condensate or other heating medium may be left in the coils and leak from their fittings. Effective air monitoring by responders early in an incident will help determine the nature of the spillage.

Glossary

AAR Association of American Railroads
ACGIH American Conference of Government and Industrial Hygienists
AIT Auto-ignition temperature. The lowest temperature at which a liquid will ignite without an independant source of combustion.
APF Assigned protection factor
APR Air-purifying respirator
ASME American Society of Metallurgical Engineers

BLEVE Boiling-liquid, expanding-vapor explosion
Buddy system A system whereby two people are assigned responsibility for each other. The system requires that buddies maintain visual contact with each other at all times.

CAER Community awareness and emergency response
CANUTEC Transport Canada's emergency centre
CCPA Canadian Chemical Producers Association
CERCLA Comprehensive Environmental Response Compensation, U.S. Liability Act
CFR Code of Federal Regulations (U.S.)
29 CFR 1910.120 Title 29 of the U.S. Code of Federal Regulations. Section 120 deals with training standards for hazardous waste site workers and personnel responding to hazardous materials emergencies.
49 CFR Title 49 of the U.S. Code of Federal Regulations. This is the Department of Transportation's regulatory document.
CGA Compressed Gas Association
CGI Combustible gas indicator

CHEMNET The CMA's response network
CHEMTREC The CMA's emergency center
CHLOREP The Chlorine Emergency Plan
CMA Chemical Manufacturers Association (U.S.)
CPC Chemical Protective Clothing
CPPI Canadian Petroleum Products Institute
CRC Chemical Referral Center (U.S.)
CRZ Contamination reduction zone
CTC Canadian Transport Commission

Degradation A chemical reaction where the molecular structure of chemical protective clothing material is actually broken down by a contaminant liquid or gas.
DoT Department of Transportation (U.S.)

EPA Environmental Protection Agency (U.S.); Environmental Protection Act (Canada)
EPC Emergency Preparedness Canada
Exothermic reaction A chemical reaction that releases heat

FEMA Federal Emergency Management Agency (U.S.)
FSB Fuels Safety Branch (Canada)
FSOP Field safe operating procedures
FWPCA Federal Water Pollution Control Act (U.S.)
FZ Flammable zone

HAZCOM Hazard Communication Standard (29 CFR 1910.1200)
Hazmat An acronym for hazardous materials

IBP Initial boiling point
ICAO International Civil Aviation Organization
ICS Incident command system
IDLH Immediately dangerous to life and health
IMDG International Maritime Dangerous Goods Code

kg Kilograms
kPa Kilopascals; the SI pressure measurement unit

LC50 Lethal concentration required to effect a 50 percent kill of laboratory test animals
LCD Liquid crystal diode
LD50 Lethal dose required, by administration, to effect a 50 percent kill of laboratory test animals
LEL Lower explosive limit
LEPC Local emergency planning committee (prescribed by SARA)

LFL Lower flammable limit
LIL Lower inflammable limit
LNG Liquefied natural gas

MAC Maximum acceptable concentration
Malleability The ease with which a material may be shaped
MAP Mutual Aid Plan (Canada)
MAWP Maximum acceptable working pressure
MC Motor carrier
MCCR Ministry of Consumer and Commercial Relations (Canada)
MOE Ministry of Environment (Canadian provinces)
MSDS Material safety data sheet (Prescribed by WHMIS and HAZCOM)
MSIS Marine Safety Information System
MSST Maximum safe storage temperature
MSTT Maximum safe transport temperature

NCP National Contingency Plan
NCRIC National Chemical Response and Information Center, a CMA initiative
NFPA National Fire Protection Association
NIOSH National Institute of Occupational Safety and Health (U.S.)
NRC National Response Center (U.S.)
NRT National Response Team (U.S.)
NTA National Transportation Agency (Canada)

OHMTADS Oil and Hazardous Materials Technical Assistance Data System (U.S.)
OHSA Occupational Health and Safety Act (Canada)
OPA Ontario Petroleum Association (Canada)
OSC On-scene commander
OSHA Occupational Safety and Health Administration (U.S.)

PACE Petroleum Association for the Conservation of the Canadian Environment (now part of CPPI)
Packing group An indication of the inherent level of danger of a product. Packing group I indicates the most dangerous type of a substance and packing group III the least. (Canada)
PAPR Powered air-purifying respirator
PCB Polychlorinated biphenyl
PEL Permissible exposure limit (OSHA)
Permeation The diffusion of a contaminant liquid or vapor through the material of chemical protective clothing. This action takes place at the molecular level.
PF Protection factor
PGAC Petroleum Gas Advisory Council (Canada)
PIN Product identification number (Canada)

PPM Parts per million
PSI Pounds per square inch
PSTFS Pollution Spill Trajectory Forecast System (U.S.)

RCRA Resource Conservation and Recovery Act (U.S.)
REL Recommended exposure limit (NIOSH)
RRC Regional Response Centre (of TEAP)
RRT Regional Response Team (U.S.)

SAE Swedish Association of Engineers
SAR Supplied-air respirator
SARA Superfund Amendments and Reauthorization Act (U.S.)
SCBA Self-contained breathing apparatus
SCDI Serious chemical distribution incident
SOP Standard operating procedures
STEL Short-term exposure limit (ACGIH)

TDGA Transportation of Dangerous Goods Act and Regulations (Canada)
TEAP Transportation Emergency Assistance Plan; the CCPA's response network.
TLV—C Threshold limit value—ceiling (ACGIH)
TLV—STEL Threshold limit value—short-term exposure limit (ACGIH)
TLV—TWA Threshold limit value—time-weighted average (ACGIH)
TRANSCAER Transportation Community Awareness and Emergency Response
TWA Time-weighted average

UEL Upper explosive limit
UFL Upper flammable limit
UIL Upper inflammable limit
U.N. United Nations
U.S. United States
USCG U.S. Coast Guard
USEPA U.S. Environmental Protection Agency

Volatility The tendency of a liquid to change to a vapor

WHMIS Workplace Hazardous Materials Information System (Canada)

Index

29 CFR 1910.120, 15
29 CFR 1910.1200, 15
49 CFR, 92
49 CFR 173.33, 158

Absorbents, 138
Acids, 82
Acute toxicity, 84
Admission valve for SCBA, 37
Adsorbents, 138
Air monitoring, 148
    for LEL, 149
    patterns, 150
    protocols, 109
    safety, 151
Air monitoring instruments, 151
    colorimetric detection tubes, 155
    combustible gas indicator (CGI), 153
    oxygen, 151
    photo ionization detector, 157
Air Purifying Respirators (APR), 45
    limiting factors, 45
Air reactivity, 83
Alergenic reactions, 91
Alkalines, 82
Anoxia, 89
Asphyxiants, chemical, 90
Asphyxiants, non-chemical, 89
Assessment, of incidents, 105
Association of American Railroads (AAR), 92

Atmospheres, 73
    inert, 74
    too lean, 73
    too rich, 73
Auto Ignition Temperature (AIT), 70

Back-Up, for entry team, 112
Bases, 82
Boiling Liquid Expanding Vapor Explosion (BLEVE), 71, 74
Boiling point, 69
Bolster, rail cars, 171
Boom, 127
    alternates to mechanical, 130
    entrainment, 130
    mechanical, 127
Breakthrough time, for chemical protective clothing materials, 57
Buddy system, 112
By-pass valve (SCBA), 40

Canadian Chemical Producers Association (CCPA), 18
Canadian Coast Guard, 17
Canadian Petroleum Producers Institute (CPPI), 18
Canadian Transport Commission (CTC), 92
CANUTEC, 17, 100
Cardiotoxins, 91

Catchbasin protection, 122
Ceiling, in threshold limit values, 86, 88
Chemical asphyxiants, 90
Chemical Manufacturers Association (CMA), 20
Chemical protective clothing, 48
 breakthrough time, 57
 compatibility, 54
 decontamination, 54, 60
 degradation, 57
 heat stress, 59
 permeation, 57
 physiological factors, 58
 precautions in use, 58
Chemical Referral Centre (CRC), 23
CHEMNET, 21
CHEMTREC, 21, 100
CHLOREP, 23
Chronic toxicity, 84
Cold zone, 112
Colorimetric detection tubes, 155
Combustible Gas Indicator (CGI), 153
Community Awareness and Emergency Response (CAER), 24
Compatibility of chemical protective clothing, 54
Containment techniques, 119
 dam, 121
 in ditches, 119
 in streams, 119
 weir, 121
Contamination reduction zone, 110
Contingency planning, 1
 by Ministry of Environment (MOE), 2
 by Resource Conservation and Recovery Act (RCRA), 2
 by Superfund Ammendments and Reauthorization Act (SARA), 1
 by Transportation of Dangerous Goods Act (TDGA), 2
Control zones, 109
Corrosivity, 82
Culvert weir, 122
Cylinder duration in SCBA, 41
Cylinders leaking, 147

Danger placard, 102
Dangerous goods, nine classes, 92

Decontamination, 61
 chemical methods, 64
 chemical protective clothing, 54, 60
 emergency, 66
 physical methods, 62
 site set up, 61
 time allowance in SCBA, 40
Degradation, of chemical protective clothing materials, 57
Dense product spills, 128
Density, vapors, 71
Disk skimmers, 135
Donning level B chemical protective clothing, 52
Donning procedure, SCBA, 41
Drum patch kit, 146
Drums, for temporary storage, 139
Drums, patching leaking, 140
Dyking, 116

Emergency decontamination, 66
Entry team back up, 112
Environment Canada, 17
Equipment, drum patching, 146
Equipment list, for emergency response outfit, 14
Evacuation plan, 4
Exclusion zone, 109, 112
Explosion, BLEVE, 71
Explosions, 71
Explosive limits, 71
Explosive zone, 73
Explosives, transport class, 93

Federal Emergency Management Agency (FEMA), 18
Federal motor carrier safety regulations, 158
Fire, 103
Fire Point, 70
First approach to an incident, 112
Fit test, 42
Flammability, gases, 70
Flammability, liquids, 68
Flammable limits, of gases, 71
Flammable liquids, transport class, 96
Flammable range, 73
Flammable solids, transport class, 97
Flash point, 69

Index   191

Gases, flammability, 70
Gases, transport class, 94

Hatch cone device, 162
Hazardous materials, nine classes, 92
Heat stress, 59
Hepatic poisons, 90
Hose contamination, supplied air respirators, 44
Hot Zone, 109

IDLH considerations, supplied air respirators, 44
Immediately Dangerous to Life and Health (IDLH), 86, 88
Incident assessment, 105
Incident Command System (ICS), 3
Incident control zones, 109
Incident, first approach, 112
Inert, 74
Infectious substances, transport class, 9
Ingestion, of toxins, 85
Inhalation, of toxins, 85
Injection, of toxins, 85
Instruments, air monitoring, 151
    colorimetric detection tubes, 155
    combustible gas indicator (CGI), 153
    oxygen, 151
    photo ionization detector, 157
Intermodal tank containers, 179
Interceptor trenches, 117
International Air Transport Authority (IATA), 92
International Civil Aviation Organization (ICAO), 92
Irritants, 88

Lethal concentration, 87
Lethal dose, 87
Levels of protection, 48
    A, 48
    B, 51
    C, 53
    D, 54
Limiting factors, air purifying respirators, 45
Line length, in supplied air respirators, 43, 44

Line pressure in supplied air respirators, 42
Liquids, flammability, 68
Low level alarm, on SCBA, 40
Lower Explosive Limit (LEL), 71
Lower explosive limit, significance in monitoring at site, 149
Lower Flammable Limit (LFL), 71
Lower Inflammable Limit (LIL), 71
Lugger boxes, for temporary storage, 139

Maximum Safe Storage Temperature (MSST), 98
MC 306/406 tank truck, 160
MC 307/407 tank truck, 163
MC 312/412 tank truck, 164
MC 331 tank truck, 165
Mechanical boom, 127
Ministries of Environment, 17
Miscellaneous dangerous goods, transport class, 101

National Chemical Response and Information Centre (NCRIC), 20
National Fire Protection Association (NFPA) 472, 16
National Institutes for Occupational Safety and Health (NIOSH), 87
National Transportation Agency, 17
Nephrotic poisons, 90
Nerve poisons, 90
Neurotoxins, 91
Neutral, 82
Non-chemical asphyxiants, 89
Non-pressure rail tank cars, 177
Non-soluble product spills, 127

Occupational Safety and Health Administration (OSHA), 15, 16
Organic peroxides, transport class, 98
Oxidizers, transport class, 98
Oxygen analyzers, 151
Oxygen considerations, supplied air respirators, 44
Oxygen deficiency, 91
Oxygen enrichment, 92

Patching leaking drums, 140
Permeability, of soil, 115
Permeation, 57
Permissible Exposure Limit (PEL), 86, 88
PERT, 23
Petroleum Association for the Conservation of the Canadian Environment (PACE), 18
pH, 82, 101
Photo Ionization Detector (PID), 157
Pits, for temporary storage, 140
Poison, 85
Poisons, 90
    hepatic, 90
    nephrotic, 90
    nerve, 90
    systemic, 90
    transport class, 99
Polymerization, 83
Portable tanks, for temporary storage, 139
Positive pressure, in air supplied respirators, 39
Powered Air Purifying Respirators (PAPR), 45
Pressure tank cars, 176
Protection factors, 46, 47
Protective clothing, 48
Pulmonary edema, 88

Radioactives, transport class, 100
Rail car bolsters, 171
Rail car trucks, 171
Rail tank car types, 176
    general description, 171
    non-pressure, 177
    pressure, 176
Reactivity, 83
    air, 83
    polymerization, 83
    self, 83
    water, 83
Recommended Exposure Limit (REL), 87, 88
Reporting forms, 9
Risk categorization, 107
Rope skimmers, 136
Routes of entry, toxicity, 85

Safety features, tank trucks, 158
Safety, in air monitoring, 151
SCBA, 37
    by-pass valve, 40
    cylinder duration, 41
    donning procedure, 41
    low level alarm, 40
SCDI, 23
Self Accelerating Decomposition Test (SADT), 84
Self Contained Breathing Apparatus (SCBA), 37
Self reactivity, 83
Sensitization, 91
Sewer invasion, 115
Sewer protection, 122
Short Term Exposure Limit (STEL), 86, 88
Skimmers, 132
    disk, 135
    rope, 136
    suction, 136
    weir, 133
Skin absorbtion, of toxins, 85
Slips, trips and falls, as incident risks, 104
Soil permeability, 115
Soluble product spills, 124
Sorbents, 138
Sorbents, in containment systems, 121
Spill management, 115
    dense products, 128
    non-soluble products, 127
    soluble products, 123
    soluble products, stream diversion, 125
    to water, 123
Stream diversion, 125
Suction skimmers, 136
Supplied Air Respirator (SAR), 41, 42
    hose contamination, 44
    IDLH considerations, 44
    line length, 43, 44
    line pressure, 42
Synergism, 91
Systemic poisons, 90

Tank truck types, 159
    MC 306/406, 160
    MC 307/407, 163
    MC 312/412, 164
    MC 331, 165
Tank truck, safety features, 158

Index   193

Tank truck, valving arrangements, 167
Terrain, as an incident risk, 104
Threshold Limit Value (TLV), 86, 88
    ceiling (C), 86, 88
    short term exposure limit (STEL), 86, 88
    time weighted average (TWA), 86, 88
Too lean, 73
Too rich, 73
Toxicity, 84
    acute, 84
    chronic, 84
    routes of entry, 85
Toxin routes, 85
    cardic system, 91
    definition of, 84
    ingestion, 85
    inhalation, 85
    injection, 85
    neural system, 91
    skin absorbtion, 85
Training requirements, 14
    for spill response, 15
    in transportation legislation, 19
    under 29 CFR 1910.120, 15
    under 29 CFR 1910.1200, 15
    under HAZCOM, 15
    under NFPA 472, 16
    under WHMIS, 15
TRANSCAER, 24
Transportation Emergency Assistance Plan (TEAP), 18
Transportation of Dangerous Goods Act (TDGA), 92

Trenching, 117
Trucks, rail cars, 171

United States Coast Guard (USCG), 20
United States Environmental Protection Agency (USEPA), 18
Upper Explosive Limit (UEL), 71
Upper Flammable Limit (UFL), 71
Upper Inflammable Limit (UIL), 71

Valving arrangements for tank trucks, 167
Vapor density, 71

Warm zone, 110
Water reactivity, 83
Water spills, 123
Water table, 115
Weir skimmers, 133
Weir, containment, 121
Weir, culvert, 122

Zones, for incident control
    cold, 112
    contamination reduction, 110
    exclusion, 109
    hot, 109
    support, 112
    warm, 110